O&B
MATHS BANK
1

K. J. Dallison M.A.
J.P. Rigby B.Sc.

Oliver & Boyd

Illustrated by David Brogan and Tom Reid

OLIVER & BOYD
Robert Stevenson House
1–3 Baxter's Place
Leith Walk
Edinburgh EH1 3BB
A Division of Longman Group Ltd.

ISBN 0 05 003154 6

First published 1978
Sixth impression 1984

Printed in Hong Kong by
Wilture Printing Co., Ltd.

230/85

O&B MATHS BANK 1

Contents

Foreword

I first came across the authors' enthusiasm as teachers ten years ago, when they founded our local mathematics association for sixth formers and invited me to become its first president.

All mathematics teachers who use one of the newer types of school syllabus will recognise the usefulness of these books. The introduction of any new syllabus should always be backed up by a wealth of problems, not only for the teacher's benefit, but also to provide side paths along which the pupils can explore and play. This is especially true in the case of the more gifted, and especially in the mixed ability classes of today, in which the more gifted must necessarily be expected to work more on their own.

Learning (that is assimilating and memorising) any syllabus can be a tiresome business, and in mathematics at any level there are always opportunities for indulging in the more challenging and exciting tasks of discovery and creativity as well. These require reflection alone, and some form of guidance. There is no better guidance than well chosen problems, that appeal to the intuition and focus the imagination, and through which the student can recreate his or her own mathematics. Such self-discovery leads to a much deeper understanding, and a confidence in the subject, which the student will never forget, and upon which he or she can build further.

E. C. ZEEMAN
University of Warwick
August 1977

Preface

These books are rather different from the usual mathematics texts in that they contain almost no teaching material. They do, however, contain a wealth of questions, covering the modern and traditional mathematics required by 'O' level and C.S.E. syllabuses in modern mathematics.

Maths Bank can be used to supplement any existing course in modern mathematics but it can also be used as a course book in its own right, leaving the teacher free to instruct in his own way.

The questions are designed to cater for a wide range of ability: each section begins with easier questions; harder and deliberately wordy questions are starred.

The authors wish to express their gratitude to Miss P. M. Southern, Mr E. P. Willin and other past and present members of the mathematics staff at Rugby High School for writing questions and supplying answers; to Mr M. E. Wardle, head of the Department of Mathematics at Coventry College of Education for acting as adviser on difficult points; and to Miss D. M. Linsley, former headmistress of Rugby High School, without whose foresight in allowing the school to change to modern mathematics in 1963 these books would never have been written.

K. J. DALLISON
J. P. RIGBY
Rugby High School
1978

1 Test Your Basic Skills

1A

1 Add together:

a) 24, 37, 5 and 341 *b*) 9853, 16 and 80
c) 2164, 11, 345 and 8 *d*) 999999 and 9999.

2 Subtract:

a) 57 from 82 *b*) 96 from 111 *c*) 13 from 21
d) 182 from 243 *e*) 799 from 812

3 Multiply:

a) 295 by 8 *b*) 371 by 7 *c*) 48 by 11
d) 316 by 5 *e*) 430 by 10

4 Divide:

a) 2429 by 7 *b*) 4866 by 6 *c*) 1331 by 11
d) 3456 by 9 *e*) 4096 by 8

5 If I bought 2 pencils which cost 7p each, a book which cost 45p and a rubber which cost 12p, how much change should I get from £1?

6 John had a bag of sweets and ate four of them himself. He gave away half of those which were left, and then found when he had eaten another one that there were only three left. How many were in the bag to begin with?

7 Jane's father gave her 10p for every examination which she passed, and took back 5p for every one she failed. If she passed six subjects out of the ten which she sat, how much money did Jane receive?

8 Mr Green wants to put a path round his pond, which is rectangular and measures 2 m × 3·5 m. The path is to be 0·5 m wide and the paving slabs are 0·5 m square. How many slabs does Mr. Green have to buy?

1B

1 Find the sum of these numbers:

a) 1·62, 13·5, 0·071, 1·008 and 4·07
b) 13·123, 0·456, 7·369, 11·1 and 4·77
c) 102·5, 3·126, 44·2, 0·066 and 7·399

2 Find the difference between:

a) 6028 and 4169 *b*) 318 and 477 *c*) 4·19 and 2·33
d) 28·22 and 37·145 *e*) 162·4 and 3·166

3 Find the value of:

a) $16{\cdot}87 \times 11$ b) $22{\cdot}3 \times 7$ c) $341{\cdot}12 \times 9$
d) $4{\cdot}166 \times 8$ e) $0{\cdot}259 \times 12$

4 Find the value of:

a) $2{\cdot}585 \div 11$ b) $4{\cdot}34 \div 7$ c) $5{\cdot}988 \div 12$
d) $37{\cdot}26 \div 6$ e) $205{\cdot}70 \div 5$

5 What is the least number which must be added to 37 to make it exactly divisible by 5?

6 Find the total cost of 15 oranges at 7p each, 3 peaches at 9p each and a box of dates at 48p.

7 Ann has £1.25 and Betty has 58p. If Ann spends half her money, which girl now has the most and how much more has she than the other girl?

8 Mr Birch wants to put a fence down one side of his garden which measures 50 m. To support the fence he has to put in a post every 2 m. How many posts does he have to buy?

9 A rectangular block is made from 2 cm cubes. If the finished block is 6 cm by 8 cm by 4 cm, how many 2 cm cubes are there in it?

1C

1 Find the value of:

a) $144{\cdot}5 + 2{\cdot}96 - 14{\cdot}07$ c) $144{\cdot}5 - 2{\cdot}96 + 14{\cdot}07$
b) $144{\cdot}5 - 2{\cdot}96 - 14{\cdot}07$ d) $144{\cdot}5 + 2{\cdot}96 + 14{\cdot}07$

Add together the answers to *a*), *b*), *c*) and *d*) and divide the result by 4. Can you explain your answer?

2 Find the products of the following:

a) $2{\cdot}071 \times 1{\cdot}2$ b) $4{\cdot}96 \times 2{\cdot}3$ c) $5{\cdot}2 \times 3{\cdot}3$ d) $1{\cdot}99 \times 5{\cdot}4$

3 Divide:

a) £1.65 by 5 b) £4.86 by 6 c) £2.94 by 7 d) £11.28 by 12

4 A family consists of one girl and three boys. The girl is the eldest and her pocket money is the same as that of the three boys put together. If the total of all four is £2.40, how much does the girl receive?

The two youngest brothers have the same amount as one another but the elder one has twice as much as either. How much do the boys each have?

5 Arrange these numbers in numerical order, starting with the smallest:

1·01, 2·001, 0·005, 1·06, 0·024.

6 A train is scheduled to leave one station at 0824 and arrive at the next one at 0908. How many minutes should the journey take?

7 What is the total of the ages of 5 children aged 5 yr 2 m, 6 yr 8 m, 6 yr 9 m, 7 yr 10 m and 8 yr 1 m?

8 The diagram shows a cube of side 5 cm. What is *a*) the area of one face, *b*) the total area of the six faces, *c*) the length of one edge, *d*) the total length of all the edges?

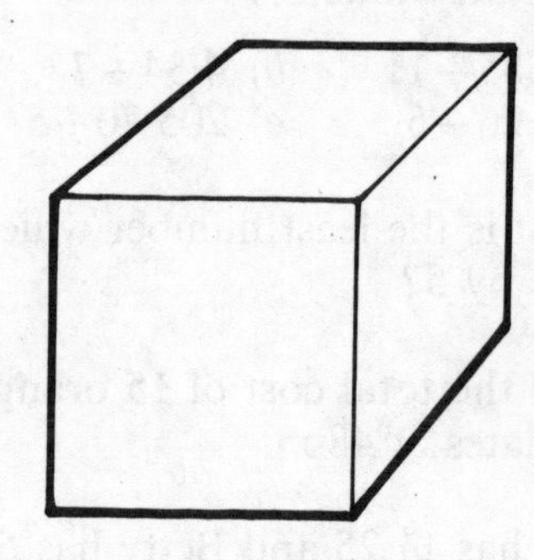

1D

1 Find the missing numbers in the following sums:

a)
$$\begin{array}{r} 164 \\ +\,2{*}8 \\ \hline {*}22 \end{array}$$

b)
$$\begin{array}{r} 4{\cdot}09 \\ +\,2{\cdot}{*}{*} \\ \hline 7{\cdot}01 \end{array}$$

c)
$$\begin{array}{r} 10{\cdot}480 \\ +\;\;{*}{\cdot}{*}{*}{*} \\ \hline 12{\cdot}962 \end{array}$$

2 If 18 bars of chocolate cost £2.34, what would 3 cost?

3 Divide *a*) 528 by 11 *b*) 528 by 33.

4 Clothes pegs are sold on cards of 12 at 9p for each card. Zoey wants to buy at least 30 pegs. How much will it cost her?

5 A coffee percolator makes 1·5 litres of coffee. How many cups each holding $\frac{1}{8}$ litre can be filled from the percolator?

6 How many bushes will be needed to make a hedge 156 m long if the bushes are planted at intervals of 0·75 m?

7 The average weight of 3 girls is 40 kg. What is their total weight?

8 Some 2 cm cubes are put together as shown in the diagram, to make a large cube.

a) How many small cubes are there?
b) If the large cube is on a table, how many of the small ones are not visible?
c) How many small cubes must be removed to leave a cube of a different size? (Two possible answers.)

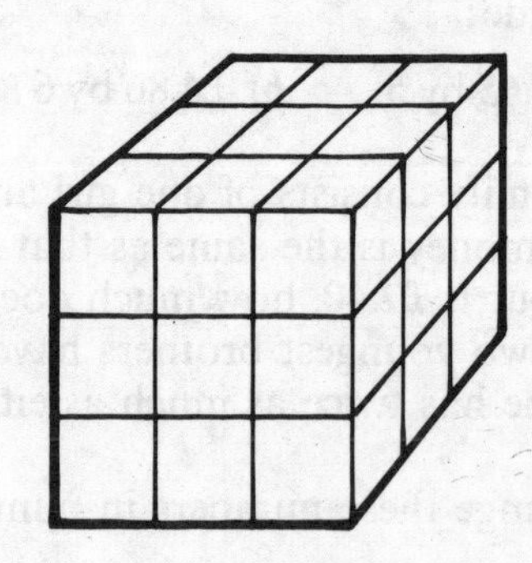

1E

1 $2{\cdot}6+1{\cdot}8-2{\cdot}3+2{\cdot}2-1{\cdot}7$

2 *a*) What must be added to 36·5 to make 42·3?
b) Two numbers added together make 12·4. One number is 3·9. What is the other?
c) What must be taken from 16·7 to leave 4·8?

3 Eight ice-creams of the same kind cost 72p. What does one cost?

4 If one book costs £2.25, what will be the total cost of 32 similar books?

5 How many pieces of string, each 3·5 m long, can be cut from a length of 17·5 m?

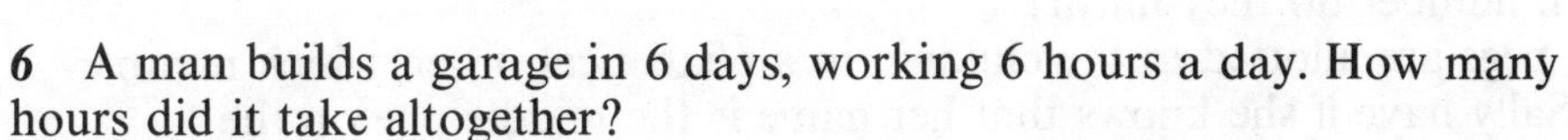

6 A man builds a garage in 6 days, working 6 hours a day. How many hours did it take altogether?
If he had had to finish in 4 days, how many hours a day would he have needed to work?

7 Find the total amount of money taken at a scout concert if 240 tickets were sold altogether: 100 at 75p each and the rest at 50p each.

8 The diagram shows a cube which is cut into smaller cubes. The outside is painted.

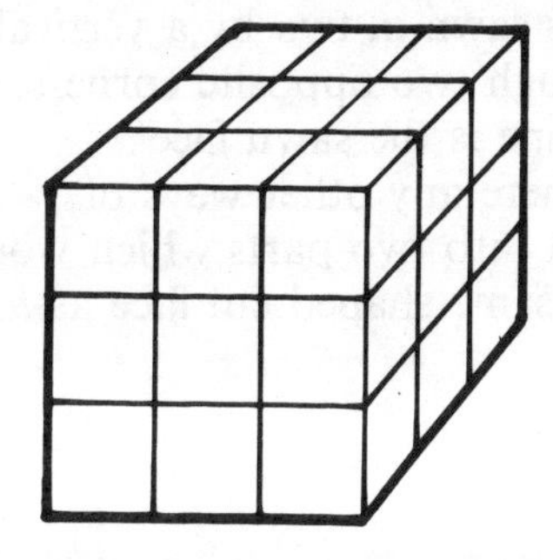

a) How many cubes have 3 painted faces?
b) How many have 2 and only 2 painted faces?
c) How many have 1 and only 1 painted face?

1F

1 Add 7p, 25p, 66p, £1.03 and £1.29.

2 The cost of a bus ride is 36p. How much will it cost for a family: mother, father and 3 children, if all the children travel for half fare?

3 Find the cost of 2·5 m of material at £3.60 per metre. How many metres could be bought for £13.50?

4 There are 48 m of mending tape on a roll. If a librarian uses 16 cm in repairing each one of a set of books, what is the highest number of books which he can repair from one roll?

5 A train averages 110 km per hour while it is moving. How long will it take to travel 165 km?

If it has two 3-minute stops during a journey of this length and starts at 0848, at what time should it complete its journey?

6 Andrew saves 25p each week. How much will he save in one year? If he continues saving at the same rate, how long will it take him to save £52?

7 A class of 33 children are given cloakroom pegs, starting at number 142. At which number do they finish?

If the pegs are allotted to the children in alphabetical order, which number should Sally have if she knows that her name is the middle one on the register?

8 The diagram shows a solid pyramid on a horizontal square base.

a) If it is sawn in two by a horizontal cut, what shape is the sawn face?

b) If it is sawn in two by a vertical cut through two opposite corners, what shape is the sawn face?

c) Are there any other ways of cutting it into two parts which would give the same shaped cut face as *b*)? Discuss.

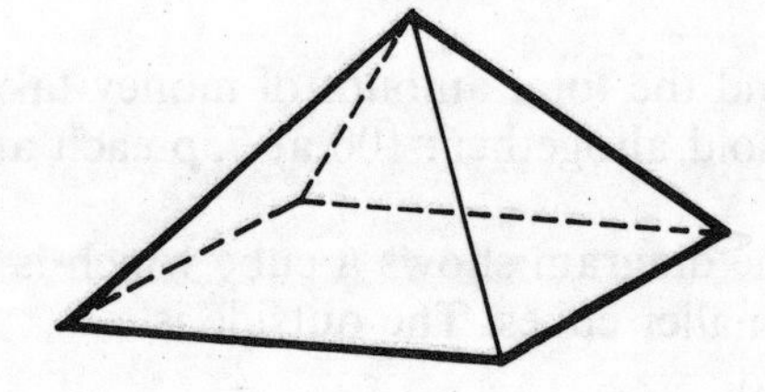

2 Sets

2A

A set is a well defined collection of objects. Do the following groups of people or things satisfy this definition? If not, say why not.

1 {Tall girls}
2 {Boys over 2 m tall}
3 {Old ladies}
4 {Sweet smells}
5 {Even numbers}
6 {Lucky numbers}
7 {Fat boys}
8 {Girls weighing 40 kg or less}
9 {Girls with brown hair}
10 {Boys with long hair}
11 {Athletic boys}
12 {Boys who play cricket}
13 {Girls who play hockey}
14 {A cat, a cabbage, a beef steak}
15 {Men over 90 years old}
16 {Bankrupts}
17 {Squares}
18 {Quadrilaterals}
19 {Tennis balls}
20 {Round objects}

2B

In 2B onwards 'numbers' is to be taken as 'positive integers' unless otherwise stated.

1 Write the members of these sets in full:

A = {Vowels}
B = {Numbers less than 10}
C = {Odd numbers less than 8}
D = {Numbers between 5 and 10}
E = {Days of the week}
F = {Months of the year beginning with J}
G = {Pupils in your class whose birthday is in May}
H = {Letters in the word SYNONYM}
I = {British coins in use today}
J = {Your brothers and sisters}

2 Describe these sets:

$A = \{A, B, C, D, E\}$

$B = \{5, 6, 7, 8, 9, 10\}$

$C = \{2, 4, 6, 8\}$

$D =$ {March, May}

$E =$ {Thursday, Friday, Saturday}

$F =$ {North, south, east, west}

$G =$ {Red, white, blue}

$H =$ {Clubs, hearts, diamonds, spades}

$I =$ {May, June, July, August}

$J =$ {Triangles, quadrilaterals, pentagons}

3 Complete the following by writing the sign '$\in$', which means 'is a member of the set', e.g. a, e, o $\in$ {vowels}.

a) P, Q, R, S
b) Monday, Tuesday, Friday
c) Crocus, tulip, snowdrop
d) Milligrams, grams, kilograms
e) 2, 3, 4
f) 4, 8, 12
g) Amber, red, green
h) Equilateral, isosceles, scalene
i) Ford, Vauxhall, Austin
j) Hereford, Coventry, Ely

4 In each of the following there is an 'odd man out'. Using the sign '$\notin$' which means 'is not a member of the set', write your answer like this: apple $\notin$ {vegetables}.

a) Pine, oak, fir, spruce
b) Chair, stool, table, sofa
c) 10, 15, 17, 20
d) Sunday, January, Tuesday, Wednesday
e) Millimetre, centimetre, kilogram, metre
f) A, B, E, I
g) 1, 2, 4, 9
h) Wool, silk, cotton, nylon
i) A, M, R, X
j) Square, triangle, rectangle, parallelogram

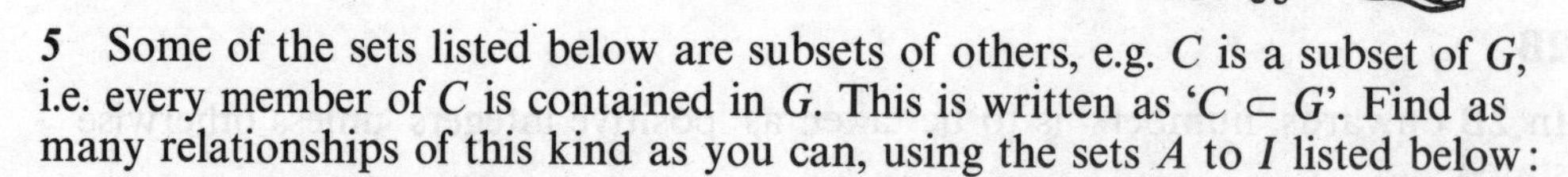

5 Some of the sets listed below are subsets of others, e.g. C is a subset of G, i.e. every member of C is contained in G. This is written as '$C \subset G$'. Find as many relationships of this kind as you can, using the sets A to I listed below:

A = {Numbers less than 20}

$B =$ {Letters of the alphabet}

$C =$ {Trees}

$D =$ {Wild flowers}

$E =$ {A, E, I, O, U}

$F = \{12, 14, 16\}$

$G =$ {Things that grow}

$H =$ {Even numbers between 10 and 20}

$I =$ {Odd numbers between 6 and 12}

Which of the above sets is ill-defined and therefore (strictly) not a set?

6 $A = \{3, 4, 5, 6, 7\}$ $D = \{\text{Integers from 1 to 20}\}$

$B = \{2, 6, 10, 14\}$ $E = \{\text{Integers from 21 to 30}\}$

$C = \{\text{Odd numbers less than 15}\}$ $F = \{\text{Odd numbers from 3 to 7}\}$

With these sets listed above, write whether the following statements are TRUE or FALSE:

a) $3 \in A$
b) $3 \in B$
c) $4 \notin B$
d) $A \subset B$
e) $A \subset C$
f) $C \subset D$
g) $D \subset B$
h) $11 \in C$
i) $11 \in D$
j) $B \subset D$

7 Write down $n(A)$ for each of the following. $n(A)$ is the 'power' of set A, i.e. the number of items in set A.

a) $A = \{\text{Even numbers less than 20}\}$

b) $A = \{\text{Odd numbers less than 20}\}$

c) $A = \{\text{Days of the week}\}$

d) $A = \{\text{British coins in use today}\}$

e) $A = \{\text{Factors of 6}\}$

f) $A = \{\text{Multiples of 3 between 10 and 20}\}$

g) $A = \{\text{Seasons of the year}\}$

h) $A = \{\text{Fingers on your left hand}\}$

i) $A = \{\text{Letters in your name}\}$

j) $A = \{\text{Numbers less than 50 containing the figure 1}\}$

8 List the members of the following sets in full.
Consider only whole numbers:

$A = \{x: 0 < x < 8\}$ $F = \{a: 40 < a < 48\}$

$B = \{x: 5 < x \leqslant 10\}$ $G = \{2a: 1 < a < 5\}$

$C = \{x: 12 \leqslant x \leqslant 20\}$ $H = \{5a: 3 \leqslant a \leqslant 7\}$

$D = \{y: 2 \leqslant y < 4\}$ $I = \{x+y: 0 < x < 4 \text{ and } 2 < y < 5\}$

$E = \{y: 15 \leqslant y \leqslant 21\}$ $J = \{a+b: 2 < a < 6 \text{ and } 10 < b < 14\}$

9 Write down 4 members of the following sets (y and x are integers i.e. positive or negative whole numbers or zero).

a) $\{y: y = x+3\}$

b) $\{y: y = 2x-4\}$

c) $\{y: y = x+5 \text{ and } x \text{ is negative}\}$

d) $\{y: y = 3x-11 \text{ and } x > 20\}$

e) $\{y: y > 10\}$

f) $\{y: y < -3\}$

g) $\{y: 2 \leqslant y \leqslant 5\}$

h) $\{y: y = 2x+4 \text{ and } 1 < x < 6\}$

i) $\{y: y = 3x-6 \text{ and } 4 > x \geqslant 0\}$

j) $\{u: u-v = 6 \text{ and } v > 2\}$

10 12 young people form a youth club: their names and ages are as follows:

Arthur (18) and his sister Jane (17)
George (17) and his sister Joan (16)
Peter (16)
Susan (16) and her brother Tony (14)
Philip (15)
Carol (15) and her sister Ann (14)
Monica (13) and her twin brother Michael (13)

Define each of the following sets in 3 ways:

B = {Jane, Joan, Susan}

C = {Arthur, George, Peter}

D = {Jane, Joan, Susan, Carol, Ann, Monica}

E = {Peter, Philip}

F = {Monica, Michael}

2C

Write down $A \cap B$ for each of the following questions:

1 A = {Even numbers less than 25}

B = {Multiples of 4}

2 A = {The years of this century with a number 1 in them}

B = {The years of this century with a number 4 in them}

3 A = {Prime numbers less than 100}

$B = \{x : 40 < x < 60\}$

4 A = {Multiples of 7 less than 40}

B = {Multiples of 3 less than 40}

5 A = {Triangles}

B = {Plane figures with all sides the same length}

6 Write down $n(A \cap B)$ for questions 1–5.

Write down $A \cup B$ for each of the following questions:

7 A = {Numbers between 50 and 70 divisible by 8}

$B = \{x : x = 5n \text{ and } 10 \leqslant n \leqslant 16\}$

8 A = {Art, Mary, Jo, Peter, Bill, Fred}

B = {Mary, Peter, Bill, Thomas, Diane, Susan, Kathleen}

9 $A = \{$Wednesdays in this month$\}$

$B = \{$5th, 10th, 15th, 20th, 25th days of this month$\}$

10 $A = \{2\,\text{km}, 1000\,\text{g}, 5\,\text{m}^2\}$ $B = \{2000\,\text{m}, 1\,\text{kg}, 5000\,\text{cm}^2\}$

11 $A = \{1, 4, 7, 10, 13, 16, 19\}$ $B = \{2, 4, 6, 8, 10, 12, 14, 16, 18\}$

12 Write down $n(A \cup B)$ for questions 7–11

2D

1 $A = \{1, 3, 5, 7, 9\}$ $B = \{$Integers from 1 to 6$\}$

Draw a Venn diagram to show these two sets. Write in full $A \cap B$.

2 $Q = \{$Quadrilaterals$\}$ $T = \{$Triangles$\}$

Draw a Venn diagram to show these two sets. Describe $Q \cap T$.

3 $Q = \{$Quadrilaterals$\}$ $S = \{$Squares$\}$

Draw a Venn diagram to show these two sets. What is $Q \cup S$?

4 $P = \{$Parallelograms$\}$ $S = \{$Squares$\}$

Draw a Venn diagram to show these two sets. What is $P \cap S$?

5 $A = \{$Numbers less than 10$\}$

$B = \{$Even numbers between 5 and 15$\}$

Draw a Venn diagram to show these two sets, and write the members of the sets $A \cap B$ and $A \cup B$.

6 $P = \{x : 6 < x < 20\}$ $Q = \{y : 1 < y < 8\}$

List the members of these two sets. Draw a Venn diagram to show the relationship between them. Shade the part of the diagram which represents $P \cap Q$ and list the members of this set.

7 $X = \{$Multiples of 3 less than 20$\}$ $Y = \{$Odd numbers less than 20$\}$

Draw a Venn diagram to show the relationship between these two sets and list the sets $X \cap Y$ and $X \cup Y$.

8 $C = \{x : 0 < x < 12\}$ $D = \{y : 4 < y < 9\}$

Draw a Venn diagram to show the relationship between these two sets. Write down a relationship between C and D. List the sets $C \cup D$ and $C \cap D$. What can you say about these last two sets?

9 $L = \{\text{Vowels}\}$ $M = \{\text{The last three letters of the alphabet}\}$

Draw a Venn diagram showing L and M. What do you know about $L \cap M$?

10 $A = \{\text{Numbers between 19 and 30}\}$ $B = \{\text{Prime numbers less than 30}\}$

Draw a Venn diagram to show the relationship of these two sets.
Describe in words $A \cap B$.

11 $G = \{\text{All the members of the class who are left handed}\}$

$H = \{\text{All the members of the class who are right handed}\}$

Illustrate these two sets on a Venn diagram.
Describe in words $G \cup H$ and $G \cap H$.

12 $L = \{\text{Prime numbers which are factors of 42}\}$

$M = \{\text{Prime numbers which are factors of 30}\}$

Draw a Venn diagram to show these two sets.

List the members of *a)* $L \cap M$ *b)* $L \cup M$.

Describe in words *a)* and *b)*.

13 $\mathscr{E} = \{\text{Pupils in a mixed school}\}$ $B = \{\text{Girls in the school}\}$

$A = \{\text{Boys in the school}\}$ $C = \{\text{Pupils under 14 in the school}\}$

Draw a Venn diagram to show these sets.

14 $\mathscr{E} = \{\text{Girls in a girls' school}\}$

$A = \{\text{Girls who eat school dinner}\}$

$B = \{\text{Girls who like playing hockey}\}$

$C = \{\text{Girls who are over 14}\}$

Draw a Venn diagram and shade $A \cap B \cap C$. Describe this set in words.

2E

In each of these questions draw a Venn diagram to illustrate the relationship of the sets.

1 $A = \{\text{Numbers less than 12}\}$

$B = \{\text{Even numbers less than 12}\}$

$C = \{\text{Odd numbers less than 12}\}$

a) What is $A \cap B$? *b)* What is $B \cup C$? *c)* What is $B \cap C$?

2 $A = \{\text{Cars in the car park}\}$ $B = \{\text{Ford cars}\}$ $C = \{\text{Red cars}\}$

Describe in words *a)* $A \cap B$ *b)* $B \cap C$.

3 X = {People in this room}

Y = {Everyone who takes a size 4 in shoes}

Z = {Everyone who takes a size 6 in shoes}

Describe in words $X \cap Y$ and $X \cap Z$.

4 L = {Prime numbers less than 20}

M = {Odd numbers between 10 and 20}

N = {Multiples of 3 between 10 and 25}

What is $L \cap M \cap N$?

5 A = {Four-sided plane figures} B = {Squares} C = {Triangles}

What is $B \cap C$?

6 P = {Trees} Q = {Deciduous trees} R = {Oak, ash, elm}

What is $P \cap Q \cap R$? What is $P \cup Q \cup R$?

7 L = {Letters of the alphabet} M = {Vowels}

N = {Letters in your name}

Describe in words $M \cap N$.

8 X = {Prime factors of 6}

Y = {Prime factors of 15}

Z = {Prime factors of 21}

Describe in words $X \cap Y \cap Z$.

9 L = {Prime factors of 10}

M = {Prime factors of 15}

N = {Prime factors of 35}

Describe in words $L \cap M \cap N$ and $L \cup M \cup N$.

10 A = {The prime numbers which multiplied together make 6}

B = {The prime numbers which multiplied together make 15}

C = {The prime numbers which multiplied together make 33}

Describe in words $A \cup B \cup C$ and $A \cap B \cap C$.

11 L = {The prime numbers which multiplied together make 6}

M = {The prime numbers which multiplied together make 35}

N = {The prime numbers which multiplied together make 21}

What is $L \cap M \cap N$? Why?

12 P = {The prime numbers which multiplied together make 10}

Q = {The prime numbers which multiplied together make 210}

R = {The prime numbers which multiplied together make 110}

What is $P \cap Q \cap R$? Why? Describe $P \cup Q \cup R$.

2F Complements of Sets

1 $\mathscr{E}$ = {Flowers} A = {Flowers with a scent} B = {Spring flowers}

Describe A', B', $A \cap B$

2 $\mathscr{E}$ = {All the houses in one road}

A = {Semi-detached houses}

B = {Houses with one garage}

Describe A', B', $A' \cap B$

3 $\mathscr{E}$ = {Books}

P = {Paperback books}

Q = {Books with coloured pictures}

Describe P', Q', $P' \cap Q$

4 $\mathscr{E}$ = {Girls in a class} Y = {Girls wearing glasses}

X = {Girls wearing brown shoes} Z = {Left-handed girls}

Describe X', Y', $X \cap Z'$, $Y' \cap Z'$, $X \cup Y'$

5 $\mathscr{E}$ = {Pencils} M = {Red pencils}

L = {Unsharpened pencils} N = {Chewed pencils}

Describe L', N', $L \cap M'$, $M \cap N$, $(M \cap N)'$

6 $\mathscr{E}$ = {Trees}

A = {Deciduous trees}

B = {Oak trees}

C = {Young trees}

Write in symbols:

a) Trees which are not deciduous
b) Mature trees
c) Young oak trees
d) Deciduous trees which are not oak
e) Which of the above sets is ill-defined, and therefore (strictly) not a set?

7 $\mathscr{E}$ = {Polygons}

P = {Triangles}

Q = {Quadrilaterals}

R = {Squares}

Write in symbols:

a) Polygons with more than 3 sides
b) Quadrilaterals which are not squares

Draw a Venn diagram to show the relationship between these sets.

8 $\mathscr{E}$ = {Letters of the alphabet written in capitals}

X = {Vowels}

Y = {Letters which are made up of straight lines}

Z = {Letters in the word MATHEMATICS}

List the members of *a*) Y' *b*) $X \cap Y$ *c*) $Z \cap X'$.

State whether the following are true or false:

d) $J \in X'$ *e*) $M \in (Y \cap Z)$ *f*) $X \subset Y$.

9 If $A = \{5, 8, 16\}$ and $A' = \{3, 9, 14\}$, what is $\mathscr{E}$?
If B is also a subset of $\mathscr{E}$ and B is $\{9, 14, 16\}$ what is B'?

10 If P = {Prime numbers less than 15}
and $P' = \{1, 4, 6, 8, 9, 10, 12, 14\}$ describe the universal set in words.

11 If M = {Even numbers less than 30}
and M' = {Odd numbers less than 30} what is the universal set?
If N = {Numbers less than 20} describe N' in words.

List the members of $M \cap N'$ and $M' \cap N'$.

12 If $A = \{a, g, m\}$ and $A' = \{c, j\}$ what is $\mathscr{E}$?
If $B \subset A$, list the possible sets B.
If $F \subset A'$, list the possible sets F.
Assume that neither B nor F are ϕ.

13 Describe in words A', B', C' from 2D, question 13,

14 Write in words A', B', C' from 2D, question 14.

15 If A = {Vowels} what is A' if

a) $\mathscr{E}$ = {Letters of the alphabet}

b) $\mathscr{E}$ = {Vowels}

c) $\mathscr{E}$ = {First 12 letters of the alphabet}

16 If $A = \{\text{Odd numbers}\}$ write A and A' in full, given that

a) $\mathscr{E} = \{\text{Positive integers less than } 10\}$

b) $\mathscr{E} = \{\text{Positive integers between } 11 \text{ and } 20\}$

c) $\mathscr{E} = \{\text{Multiples of } 5 \text{ less than } 30\}$

17 If $\mathscr{E} = \{\text{Months of the year}\}$, $A = \{\text{Months of the year beginning with J}\}$, $B = \{\text{Months of the year beginning with A}\}$, $C = \{\text{Months of the year beginning with M}\}$, $D = \{\text{Months of the year with an R in their name}\}$.

a) Describe in words A', B', C', D'

b) Describe in words $A' \cap B'$ and $(A \cap B)'$.

2G Complements of Sets

1 $\mathscr{E} = \{\text{Numbers less than } 25\}$ $\quad B = \{\text{Prime numbers}\}$

$A = \{\text{Even numbers}\}$ $\quad C = \{\text{Multiples of } 3 \text{ between } 5 \text{ and } 20\}$

List the members of

a) A' *b)* B *c)* C *d)* $A \cap B$ *e)* $A' \cap C$

2 $\mathscr{E} = \{\text{Numbers between } 5 \text{ and } 25\}$ $\quad G = \{\text{Multiples of } 4\}$

$F = \{\text{Multiples of } 10\}$ $\quad H = \{\text{Even numbers}\}$

List the members of

a) F *b)* $F \cap G$ *c)* $F' \cap H$ *d)* $F \cup G'$ *e)* $H' \cap G$

3 $\mathscr{E} = \{\text{Numbers between } 8 \text{ and } 28\}$

$L = \{\text{Multiples of } 3\}$

$M = \{\text{Factors of } 72\}$

$N = \{\text{Numbers whose digits add up to } 6\}$

List the members of

a) M *b)* N *c)* $N \cap L'$ *d)* $M \cap L'$ *e)* $M \cap L$

4 $\mathscr{E} = \{\text{Letters in the words ARITHMETIC BOOKS}\}$

$W = \{\text{Letters in METRIC}\}$

$X = \{\text{Letters in MATHS}\}$

$Y = \{\text{Letters in BRACKETS}\}$

$Z = \{\text{Vowels}\}$

List the members of *a)* W' *b)* Y' *c)* $X \cap Z'$ *d)* $X' \cap Y$ *e)* $Z \cap Y$

Draw a Venn diagram to show $\mathscr{E}$, W, X and Y.

5 $\mathscr{E} = \{\text{Numbers from 1 to 20 inclusive}\}$

a) If $A' = \{1, 2, 3, 4, 5\}$, list set A.

b) If $B' = \{1, 2, 3, 5, 7, 11, 13\}$, list set B.

List the members of *c*) $A' \cap B'$ *d*) $A \cap B'$ *e*) $A \cup B$.

Show these sets in a Venn diagram.

6 $\mathscr{E} = \{\text{Numbers from 10 to 20 inclusive}\}$

a) If $X' = \{10, 12, 14, 16, 18, 20\}$, describe X in words.

b) If $Y' = \{10, 15, 20\}$, list the members of Y.

c) If $Z' = \{12, 15, 18\}$, list the members of Z.

List the members of *d*) $X \cap Z'$ *e*) $Y' \cap X'$ *f*) $X' \cap Z'$.

Describe each of these three sets in words.

7 $\mathscr{E} = \{a, c, e, j, m, r, s\}$

a) If $X' = \{a, c, e, s\}$ list the members of X.

b) If $Y = \{j, m\}$ what can you say about the relationship between X and Y?

c) If $Z = \{a, e\}$ draw a Venn diagram to show X, Y and Z.

8 If $M = \{b, d, f, g, j, k\}$ and $M' = \{c, h\}$ what is $\mathscr{E}$?

If $N = \{b, c, f, j, k\}$ list the members of N'.

List $M \cap N$ and $N \cap M'$ and draw a Venn diagram to show M and N.

9 $\mathscr{E} = \{\text{Numbers less than 20}\}$

Write in each of the following, the numbers represented by $*$.

a) $\{1, 4, 9, 12\} \cap \{*, 15\} = \{9\}$

b) $\{5, 6, 9, 10, 17, 18\} \cup \{*, 7, 11\} = \{4, 5, 6, 7, 9, 10, 11, 17, 18\}$

c) $\{6, 9, 14, 15, *\} \cap \{12, 15, 18, 19\} = \{15, 18\}$

d) $\{3, 5, 9, *, 17\} \cap \{8, 10, *, 15, 17\} = \{12, 17\}$

10 $\mathscr{E} = \{\text{Prime numbers less than 25}\}$

$A = \{\text{Prime factors of 30}\}$

$B = \{\text{Prime factors of 55}\}$

$C = \{\text{Prime factors of 130}\}$

List the members of A', B', C', $A' \cap B' \cap C'$.

Describe the set $A' \cap B' \cap C'$ in words.

Draw a Venn diagram to show A, B and C.

11 Look at 2B, question 2.

Define A' given that $\mathscr{E} = \{\text{the first ten letters of the alphabet}\}$
Define B', C', $(B \cap C)'$ and $B' \cap C'$ given that $\mathscr{E} = \{\text{positive integers 1 to 10}\}$.
Define D', I' given that $\mathscr{E} = \{\text{the first eight months of the year}\}$
Define J' given that $\mathscr{E} = \{\text{polygons with eight sides or less}\}$
Define H' given that $\mathscr{E} = \{\text{a standard set of playing cards (no jokers)}\}$.

12 In 2C, question 1, if $\mathscr{E} = \{\text{positive integers less than 30}\}$ write down A', B', $A' \cap B'$ and $(A \cap B)'$.

13 In 2C, question 3, if $\mathscr{E} = \{\text{positive integers less than 100}\}$ describe in words A', B', and $A' \cap B'$.

14 In 2C, question 5, if $\mathscr{E} = \{\text{polygons}\}$ describe in words A', B', $(A \cap B)'$ and $A' \cap B'$.

15 If A is any set whatever, describe in words and in symbols $A \cap A'$ and $A \cup A'$.

16 If $A = \{\text{All roads leading to Gloucester}\}$

$B = \{\text{All roads that are flooded}\}$

what do you deduce from the following:

a) $A \cap B = \phi$ *b)* $A \cap B = A$ *c)* $n(A \cap B) = 2$

2H

By drawing Venn diagrams, find the answers to the following:

1 Of 30 girls in a class, 16 played hockey and 20 played netball. If they all played at least one game, find how many played both hockey and netball.

2 On a Sunday morning a boy delivered papers to 50 houses. He delivered 26 *Reflections* and 32 *Echos*. How many households had both papers?

3 Out of 100 people questioned about their preference for strawberry or vanilla ices, 25 ate only vanilla. 60 ate both. Assuming that they all ate ice cream, how many liked strawberry only?

4 60 people guessed correctly the contents of one or both of two bags, A and B. Twice the number of those who guessed both correctly guessed only A right, and three times the number who guessed both guessed only B correctly. How many knew what was in A and how many knew what was in B?

5 35 pupils in a class were asked if they had bacon and eggs for breakfast. Seven of them had neither, fifteen of them had both bacon and eggs and six of them had eggs only. How many had bacon only?

6 In a school 60 girls were doing languages in the sixth form:

15 were doing French only
4 were doing French and Latin only
5 were doing Latin only
2 were doing Latin and German only
6 were doing French and German only
20 were doing German only

How many girls were doing all three languages?

* *7* In a science sixth form of 30 pupils, 20 were doing Physics, 15 were doing Chemistry and 18 were doing Biology. 10 were doing both Chemistry and Biology, 13 were doing both Physics and Chemistry and 9 did all three subjects.

a) How many were doing Chemistry only?
b) How many were doing Biology and Physics only?
c) How many were doing Biology only?

8 The pupils in a school had a choice of three games in the summer term. In one class of 30, no one played all three games but 16 played tennis, and 2 of these played basketball as well while 4 played cricket as well. 5 pupils played cricket and no other game, and 6 played both cricket and basketball.

a) How many played just tennis?
b) How many played basketball?

9 50 people were asked which of three newspapers they read. 20 read the *Globe*, 23 the *News* and 25 the *Daily Times*. Of the 11 who read both the *News* and the *Times*, 3 also read the *Globe*. 10 read only the *Times*. No one read just the *Globe* and *News*.

a) How many read just the *Globe*?
b) How many read just the *News*?

10 At lunch time one day a restaurant served soup, a meat course and a pudding. The cashier kept a note of which courses were being eaten. 21 people had soup, 38 had the meat course and 20 had pudding. Two people had only soup and one had soup and pudding but no meat course. 15 people had all three courses and 4 had meat and pudding but no soup.

a) How many had just the meat course?
b) How many people altogether had at least one course?

3 Bases

3A

1 A toy scales consists of two scale pans, on one of which an article to be weighed is placed, and on the other, selected weights from a set.

a) If this set consists of five weights: 1 g, 2 g, 4 g, 8 g and 16 g, what is the maximum weight you can weigh?
b) If there were six weights, what would the sixth weight be? What would the extra two weights be in a set of eight weights?
c) With the set of five weights, are there any weights (in whole grams) less than the above maximum which it would be impossible to weigh? If so, state them.

2 The following table shows how certain weights are made up. A '1' is placed in the appropriate column if that weight is used. A shortened version of the collection of weights is written in the last column.

Total weight	*32 g*	*16 g*	*8 g*	*4 g*	*2 g*	*1 g*	*Shortened version*
9 g			1			1	1001
14 g			1	1	1		1110
24 g		1	1				11000
35 g	1				1	1	100011

a) What do the '0's signify?

b) Make a similar table and write the weights needed to balance the following total weights, writing each answer first under the column headings and then in the shortened version:

5 g 7 g 11 g 12 g 20 g 28 g 31 g 40 g 45 g 63 g

3 *a*) If some scales came with weights 1 g, 3 g, 9 g, 27 g instead of the set described in question 1, what is the maximum weight which the scales could weigh with the set of four weights?

b) How many different total weights would they weigh using any number of the four weights at a time?

c) To make it possible to weigh any total in whole grams up to 60 g how many of the specified sets of weights would be required? What would the maximum weight now be?

4 With the sets of weights you have decided you need in question 3c, complete this table in a similar way to the one in question 2:

Total weight	*27 g*	*9 g*	*3 g*	*1 g*	*Shortened version*
11 g		1		2	102
16 g		1	2	1	121

26 g 30 g 36 g 40 g 42 g 55 g 61 g 74 g 76 g 80 g

5 Another set of weights consists of three weights only: 1 g, 5 g and 25 g. How many such sets would be required to weigh any weight up to 110 g in whole grams?

a) What would the maximum weight now be?
b) Complete a table similar to those in questions 2 and 4 for these total weights:

6 g 9 g 12 g 15 g 24 g 31 g 53 g 62 g 100 g 124 g

6 Using column headings – 16's, 4's, 1's – in a similar way to those in the tables of weights, change these denary numbers to base 4 numbers:

6 10 16 23 60 37 49 21 40 56

7 *a*) If you used column headings to change denary numbers to base 7 what would the column headings be?
b) Use them to change these denary numbers to base 7:

10 15 28 39 75 91 103 112 30 70

8 4 girls – Anne, Betty, Carole and Denise – compete in a life saving test. Each girl has to carry out 4 assignments of increasing difficulty. The points awarded were 1, 2, 4 and 8 for tests 1, 2, 3 and 4 respectively. Here are the results: (P = pass; F = fail)

	Test 1	*Test 2*	*Test 3*	*Test 4*
Anne	P	P	F	P
Betty	P	F	P	P
Carole	P	P	P	F
Denise	P	F	F	P

Write down the scores of the 4 girls in binary, and find in binary the total number of points scored by the 4 girls.

9 A large cake is cut up into small pieces by the following method. It is cut into halves. Each piece is then halved. Each piece is halved again. This is done a total of 10 times.

a) Write in binary the number of pieces obtained.
b) If one piece falls on the floor at the 4th and again at the 7th cut, work out in binary how many pieces are actually obtained.
c) Change both your answers to denary.

10 A valuable book of sketches, containing 16 pages, is priced at £500. A collector haggles over the price asked. The bookseller offers him the book at 1p for the first page, 2p for the second, 4p for the third, etc. The collector agrees, and is very surprised to find that the book now costs him considerably more than before. Write down in binary the cost of the book, convert it to denary and find the actual price paid. This price is 1p less than would have been paid for the 17th page, if there had been a 17th page. Is this a general rule? Discuss.

* *11* A customer, trying to buy a horse from a dealer, was reluctant to pay the price being asked – £200. The dealer then offered to sell him the horse for 1p for the first nail in the hooves, 2p for the second nail, 4p for the next nail, etc. The customer agreed enthusiastically. There were 28 nails altogether. How much would the customer have had to pay if the dealer had made him stick to his promise?

* **12** A sheet of exercise-book paper is about 0·1 mm thick. If such a sheet is folded in two, and then in two again and so on, making altogether 24 folds, how high would the folded paper stand? Give your answer to the nearest tenth of a kilometre.

Why is the answer absurd? Try folding a sheet of exercise book paper yourself and see how far you get towards the 24 folds.

3B

Change the following binary numbers into denary numbers:

1	100	*6*	10010	* *11*	100001	* *16*	11011001
2	111	7	101011	* *12*	110001	* *17*	1111111
3	10101	*8*	110111	* *13*	101010	* *18*	10110110
4	11011	*9*	110110	* *14*	11011000	* *19*	1001001001
5	10111	*10*	111101	* *15*	1010110	* *20*	1101101010

3C

Change the following denary numbers into binary numbers:

1	11	*6*	196	* *11*	1023	* *16*	6913
2	27	7	218	* *12*	1025	* *17*	8844
3	43	*8*	431	* *13*	2147	* *18*	17 688
4	85	*9*	729	* *14*	2222	* *19*	1 000 000
5	127	*10*	946	* *15*	4096	* *20*	2 000 000

3D

Work out the following binary sums in binary. For numbers 1 to 10 check your answers by converting to denary:

1 111 + 101

2 101 + 101

3 111 + 111

4 1011 + 100

5 1100 + 1011

6 1010 + 110

7 1111 + 1011

8 1111 + 1001

9 11000 + 101

10 11011 + 110

11 101101 + 110

12 101010 + 1100

13 11010 + 1101

14 100110 + 1011

15 101 + 110 + 111

16 111 + 101 + 1011

17 1001 + 1110 + 1001

18 1101 + 1001 + 1111 + 1000

19 11 + 101 + 1101 + 11101 + 110111

20 1 + 11 + 111 + 1111 + 11111

3E

Work out the following in binary. Check your answers to the first ten by converting to denary.

Check numbers 11 to 20 by adding the two bottom lines of your sum to see if they give the first line.

1 110 − 101

2 111 − 101

3 100 − 11

4 1011 − 101

5 1100 − 11

6 10101 − 1001

7 11001 − 101

8 1101 − 1011

9 1111 − 1011

10 1001 − 111

11 10110 − 111

12 11011 − 1010

13 10111 − 10101

14 11011 − 1101

15 11010 − 1111

16 101011 − 11001

17 100100 − 11011

18 100101 − 11010

19 1000000 − 111111

20 10000000 − 1111111

3F

Find the values of the following in binary. Do a check in denary for numbers 5 to 15.

1	11×10	*6*	1101×110	* *11*	10001×111	* *16*	1001×1001
2	11×11	7	1011×101	* *12*	110000×101	* *17*	11010×100
3	101×11	*8*	10110×101	* *13*	101100×110	* *18*	1110×1010
4	101×110	*9*	11011×110	* *14*	101101×101	* *19*	10110×1101
5	111×11	*10*	1011×1101	* *15*	11011×1011	* *20*	111000×101

3G

Find the value of the following in binary. Where the division is not exact, state the remainder. Do a check in denary for numbers 5 to 15.

1	$1001 \div 11$	*6*	$1111 \div 101$	* *11*	$101010 \div 111$	* *16*	$10101 \div 1111$
2	$1101 \div 11$	7	$11111 \div 101$	* *12*	$110111 \div 101$	* *17*	$111011 \div 10101$
3	$10100 \div 100$	*8*	$10101 \div 110$	* *13*	$11101 \div 1011$	* *18*	$1011011 \div 1101$
4	$111011 \div 111$	*9*	$10100 \div 101$	* *14*	$11011 \div 1001$	* *19*	$1010101 \div 1011$
5	$10010 \div 110$	*10*	$11001 \div 101$	* *15*	$10101 \div 1101$	* *20*	$1010101 \div 10101$

3H

In this exercise all the numbers are in binary and the sign $\underset{*}{**}$ stands for $+$, $-$, $\times$ or $\div$. State which.

1	$111 \underset{*}{**} 101 = 10$	*6*	$1111 \underset{*}{**} 101 = 11$
2	$101 \underset{*}{**} 111 = 100011$	7	$1111 \underset{*}{**} 101 = 1010$
3	$11 \underset{*}{**} 11 = 1001$	*8*	$1111 \underset{*}{**} 101 = 10100$
4	$11 \underset{*}{**} 11 = 110$	*9*	$1111 \underset{*}{**} 101 = 1001011$
5	$1010 \underset{*}{**} 101 = 10$	*10*	$110001 \underset{*}{**} 1111 = 100010$

3I

1 Change these numbers which are all in base 3 to base 10:

21 201 122 1212 11022

2 Change these numbers which are all in base 5 to base 10:

23 142 204 341 1242

3 Change these numbers which are all binary numbers to base 10:

11 101 110 1011 11011

4 Change these numbers which are all in base 8 to base 10:

23 75 66 125 243

5 These numbers are all in the given base. Change them to denary:

a) 124_5 *b*) 32_7 *c*) 201_4 *d*) 1001_2
e) 312_4 *f*) 62_8 *g*) 2021_3 *h*) 1010_2
i) 222_3 *j*) 51_7 *k*) 231_4 *l*) 452_6
m) 175_8 *n*) 234_5 *o*) 2002_3 *p*) 44_6
q) 110011_2 *r*) 12011_3 *s*) 242_5 *t*) 1311_4

3J

1 Change these denary numbers into binary numbers:

6 8 13 17 32

2 Change these denary numbers into base 4:

5 8 17 25 67

3 Change these denary numbers into base 5:

5 8 11 26 78

4 Change these denary numbers into base 8:

5 8 24 42 67

5 Change these denary numbers into the given bases:

a) 25 to base 2
b) 13 to base 3
c) 33 to base 8
d) 85 to base 4
e) 63 to base 5
f) 162 to base 5
g) 72 to base 3
h) 129 to base 7
i) 48 to base 2
j) 51 to base 4
k) 252 to base 8
l) 128 to base 5
m) 68 to base 2
n) 100 to base 3
o) 75 to base 2
p) 163 to base 7
q) 525 to base 7
r) 422 to base 8
s) 128 to base 2
t) 128 to base 4

* 3K

Change the following denary numbers into binary numbers via octal:

Example:

$$\begin{array}{l} 8\,\underline{)\,347} \\ 8\,\underline{)\,43}\quad \text{r. } 3 \\ \quad\ 5\quad \text{r. } 3 \end{array}$$

$$347_{10} = 533_{8}$$

Now write each digit in your octal answer as a 'three figure group' in binary. (0 = 000 1 = 001 2 = 010 3 = 011 4 = 100 5 = 101 6 = 110 7 = 111)

The zeros must not be dropped except in front of the first figure.

We now have $347_{10} = 533_{8} = 101011011_{2}$

Another example: $98_{10} = 142_{8} = 001100010_{2}$ which may be written 1100010_{2}.

1	197	2	417	*3*	864	*4*	2143	5	6262
6	8914	7	23 461	*8*	100 000	9	200 000	*10*	1 000 000

3L

Add together these pairs of numbers in the bases given:

1 Base 3: $\begin{array}{r} 121 \\ +\ 11 \\ \hline \end{array}$ 2 Base 4: $\begin{array}{r} 233 \\ +\ 12 \\ \hline \end{array}$ 3 Base 7: $\begin{array}{r} 43 \\ +524 \\ \hline \end{array}$ 4 Base 5: $\begin{array}{r} 134 \\ +142 \\ \hline \end{array}$

5 Base 3: $\begin{array}{r} 1022 \\ +\ 122 \\ \hline \end{array}$ *6* Base 5: $\begin{array}{r} 314 \\ +244 \\ \hline \end{array}$ 7 Base 7: $\begin{array}{r} 226 \\ +354 \\ \hline \end{array}$ *8* Base 4: $\begin{array}{r} 333 \\ +212 \\ \hline \end{array}$

9 Base 8: $\begin{array}{r} 714 \\ +346 \\ \hline \end{array}$ *10* Base 9: $\begin{array}{r} 882 \\ +288 \\ \hline \end{array}$ *11* Base 7: $\begin{array}{r} 361 \\ +466 \\ \hline \end{array}$ *12* Base 3: $\begin{array}{r} 112 \\ +212 \\ \hline \end{array}$

13 Base 6: $\begin{array}{r} 524 \\ +243 \\ \hline \end{array}$ *14* Base 7: $\begin{array}{r} 262 \\ +354 \\ \hline \end{array}$ *15* Base 8: $\begin{array}{r} 562 \\ +371 \\ \hline \end{array}$ *16* Base 4: $\begin{array}{r} 233 \\ +312 \\ \hline \end{array}$

17 Base 3: $\begin{array}{r} 101 \\ +212 \\ \hline \end{array}$ *18* Base 5: $\begin{array}{r} 421 \\ +134 \\ \hline \end{array}$ *19* Base 4: $\begin{array}{r} 123 \\ +122 \\ \hline \end{array}$ *20* Base 6: $\begin{array}{r} 445 \\ +352 \\ \hline \end{array}$

3M

Work out the following subtractions in the bases given:

1 Base 4 232 −133	*2* Base 3 211 −122	*3* Base 6 454 −244	*4* Base 9 368 −172
5 Base 4 321 −123	*6* Base 8 674 −536	*7* Base 5 432 −124	*8* Base 7 524 − 32
9 Base 8 245 −136	*10* Base 6 424 − 33	*11* Base 6 352 − 44	*12* Base 5 2120 − 323
13 Base 4 312 −113	*14* Base 7 462 −363	*15* Base 6 543 −354	*16* Base 5 411 −232
17 Base 9 361 −278	*18* Base 5 1411 − 134	*19* Base 4 1321 − 333	*20* Base 8 2166 − 372

3N

Multiply the following pairs of numbers together in the bases given.

1 Base 9 21×12

2 Base 3 21×12

3 Base 5 21×12

4 Base 4 33×22

5 Base 3 11×10

6 Base 6 35×44

7 Base 7 62×26

8 Base 9 87×43

9 Base 8 63×37

10 Base 8 131×24

* *11* Base 5 221×43

* *12* Base 4 211×13

* *13* Base 3 222×22

* *14* Base 7 456×23

* *15* Base 8 447×66

* *16* Base 6 525×43

* *17* Base 4 313×21

* *18* Base 6 543×21

* *19* Base 9 777×351

* *20* Base 5 4241×314

Do a denary check on numbers 1 to 10.

3O

Work out the following giving your answers in the base stated. Where the division is not exact state the remainder.

1	Base 3	$222 \div 12$	*	*11*	Base 8	$3714 \div 62$
2	Base 6	$435 \div 34$	*	*12*	Base 7	$4416 \div 63$
3	Base 5	$234 \div 43$	*	*13*	Base 4	$2221 \div 13$
4	Base 7	$614 \div 35$	*	*14*	Base 6	$4522 \div 521$
5	Base 4	$1233 \div 23$	*	*15*	Base 5	$42424 \div 344$
6	Base 9	$766 \div 25$	*	*16*	Base 7	$25143 \div 443$
7	Base 8	$467 \div 35$	*	*17*	Base 8	$2144 \div 37$
8	Base 3	$2121 \div 12$	*	*18*	Base 3	$11011 \div 222$
9	Base 6	$4351 \div 51$	*	*19*	Base 9	$4638 \div 57$
10	Base 5	$4444 \div 23$	*	*20*	Base 4	$12333 \div 232$

Do a denary check for numbers 11 to 20.

* 3P Bicimals

Note You will find it helpful in doing this exercise to construct a table of bicimal and decimal fractions as shown below.

Bicimal	*Decimal*
0·1	$\frac{1}{2}$ or 0·5
0·01	$\frac{1}{4}$ or 0·25
. . .	. . .
0·00001	$\frac{1}{32}$ or 0·031 25

Change these bicimal fractions into decimal fractions:

1 0·1

2 0·01

3 0·11 (i.e. 0·1 + 0·01)

4 0·0001

5 0·1001 (i.e. 0·1 + 0·0001)

6 0·100

7 0·101

8 0·0011

9 0·0101

10 0·10101

Change these decimal fractions into bicimal fractions:

11 0·5

12 0·25

13 0·125

14 0·0625

15 0·5625 (i.e. 0·5 + 0·0625)

16 0·375 (i.e. 0·25 + 0·125)

17 0·625

18 0·1875

19 0·140 625

20 0·093 75

3Q Finding the Base.

In each question state the base used.

1	22+22 = 121	*11*	452−345 = 103
2	1231−123 = 1103	*12*	110234÷232 = 212
3	333×111 = 40263	*13*	333×333 = 332001
4	100122÷101 = 222	*14*	211−122 = 12
5	111+111 = 1110	*15*	232+332 = 1004
6	222+121 = 1003	*16*	1056÷22 = 43
7	112×321 = 36052	*17*	21×12 = 312
8	325−254 = 51	*18*	111+101 = 1100
9	102420÷123 = 432	*19*	321−233 = 33
10	113×231 = 26323	*20*	2155÷43 = 34

21 If you were born in the year 1001 and died in the year 3001, assuming that no one lives to be more than 130 years old, (denary scale), in what scale of notation are the dates written?

3R

Give the meaning of the sign $\overset{**}{*}$ (it is +, −, × or ÷) in questions 1 to 10.

1	Base 4 23 $\overset{**}{*}$ 32 = 121	*6*	Base 5 2210 $\overset{**}{*}$ 122 = 2033
2	Base 2 110001 $\overset{**}{*}$ 111 = 111	*7*	Base 6 34110 $\overset{**}{*}$ 234 = 123
3	Base 8 32 $\overset{**}{*}$ 23 = 756	*8*	Base 7 123366 $\overset{**}{*}$ 243 = 342
4	Base 3 121 $\overset{**}{*}$ 112 = 1010	*9*	Base 5 111 $\overset{**}{*}$ 131 = 20041
5	Base 9 654 ** 546 = 107	*10*	Base 8 3223 $\overset{**}{*}$ 516 = 2505

Fill in the blank and state the base in each of the following:

11
```
  323
 +3*3
 1302
```

12
```
   2*1
  ×233
 52603
```

13
```
   111
 +10*1
 10010
```

14
```
3*)1341
     23
```

15
```
   121
  ×1*1
 21201
```

16
```
  23*
 +432
 1111
```

17
```
3*)1024
    023
```

18
```
  1111
  −22*
   112
```

19
```
  4*3
 −225
  235
```

20
```
   111
  ×*21
 41131
```

4 Fractions

HALVES

QUARTERS

EIGHTHS

4A

Fill in the gaps in the following making each a row of equivalent fractions:

1 $\frac{1}{2} = \frac{}{4} = \frac{}{6} = \frac{}{8} = \frac{}{10} = \frac{}{12} = \frac{}{14} = \frac{}{20} = \frac{}{24}$

2 $\frac{1}{5} = \frac{}{10} = \frac{}{15} = \frac{}{20} = \frac{}{25} = \frac{}{30}$

3 $\frac{3}{5} = \frac{}{10} = \frac{}{15} = \frac{}{20} = \frac{}{25} = \frac{}{30} = \frac{}{45} = \frac{}{60}$

4 $\frac{2}{3} = \frac{}{6} = \frac{}{9} = \frac{}{12} = \frac{}{15} = \frac{}{18} = \frac{}{21} = \frac{}{24}$

5 $\frac{3}{4} = \frac{}{8} = \frac{}{12} = \frac{}{16} = \frac{}{20} = \frac{}{24} = \frac{}{28}$

6 $\frac{5}{6} = \frac{}{12} = \frac{}{18} = \frac{}{24} = \frac{}{30}$

7 $\frac{3}{7} = \frac{}{14} = \frac{}{21} = \frac{}{28} = \frac{}{35}$

8 $\frac{1}{8} = \frac{}{16} = \frac{}{24} = \frac{}{32} = \frac{}{40}$

9 $\frac{5}{8} = \frac{}{16} = \frac{}{24} = \frac{}{32} = \frac{}{40}$

10 $\frac{2}{9} = \frac{}{18} = \frac{}{27} = \frac{}{36} = \frac{}{45}$

11 $\frac{1}{12} = \frac{}{24} = \frac{}{36} = \frac{}{48} = \frac{}{60}$

12 $\frac{5}{12} = \frac{}{24} = \frac{}{36} = \frac{}{48} = \frac{}{60}$

Using the above fractions find the answers to the following:

13 $\frac{1}{2} + \frac{1}{5}$ **14** $\frac{3}{5} + \frac{2}{3}$ **15** $\frac{2}{3} + \frac{3}{4}$ **16** $\frac{3}{4} + \frac{1}{8}$

17 $\frac{5}{6} + \frac{1}{2}$ **18** $\frac{5}{8} + \frac{5}{12}$ **19** $\frac{2}{9} + \frac{5}{6}$ **20** $\frac{1}{12} + \frac{2}{9}$

21 $\frac{5}{12} + \frac{5}{8}$ **22** $\frac{1}{8} + \frac{2}{3}$ **23** $\frac{2}{3} - \frac{3}{7}$ **24** $\frac{1}{2} - \frac{1}{12}$

25 $\frac{3}{5} - \frac{5}{12}$ **26** $\frac{1}{2} - \frac{3}{7}$ **27** $\frac{5}{12} - \frac{2}{9}$ **28** $\frac{2}{3} - \frac{1}{12}$

29 $\frac{5}{6} - \frac{1}{5}$ **30** $\frac{3}{4} - \frac{3}{7}$ **31** $\frac{3}{4} - \frac{5}{12}$ **32** $\frac{3}{5} - \frac{2}{9}$

33 $\frac{1}{2} + \frac{2}{3} + \frac{5}{6}$ **34** $\frac{1}{12} + \frac{5}{6} - \frac{1}{2}$ **35** $\frac{5}{12} + \frac{5}{8} - \frac{3}{4}$

4B

1 Add *a)* $\frac{1}{2} + \frac{1}{3}$ *b)* $\frac{1}{4} + \frac{1}{3}$ *c)* $\frac{1}{4} + \frac{1}{6}$

2 Add *a)* $\frac{2}{3} + \frac{1}{2}$ *b)* $\frac{3}{4} + \frac{2}{3}$ *c)* $\frac{3}{4} + \frac{5}{6}$

3 Add *a)* $\frac{1}{8} + \frac{1}{12}$ *b)* $\frac{3}{8} + \frac{5}{12}$ *c)* $\frac{5}{8} + \frac{7}{12}$

4 Add *a)* $\frac{1}{9} + \frac{1}{15}$ *b)* $\frac{1}{15} + \frac{1}{20}$ *c)* $\frac{1}{12} + \frac{1}{18}$

5 Add *a)* $\frac{2}{9} + \frac{2}{15}$ *b)* $\frac{7}{15} + \frac{7}{20}$ *c)* $\frac{5}{12} + \frac{5}{18}$

6 Add *a)* $\frac{1}{3} + \frac{1}{9}$ *b)* $\frac{1}{4} + \frac{1}{20}$ *c)* $\frac{1}{5} + \frac{1}{15}$

7 Add *a)* $\frac{2}{3} + \frac{2}{9}$ *b)* $\frac{3}{4} + \frac{3}{20}$ *c)* $\frac{4}{5} + \frac{4}{15}$

8 Add *a)* $\frac{4}{15} + \frac{3}{20}$ *b)* $\frac{11}{21} + \frac{2}{27}$ *c)* $\frac{7}{18} + \frac{13}{24}$

9 Add *a)* $\frac{4}{5} + \frac{1}{3}$ *b)* $\frac{3}{10} + \frac{5}{6}$ *c)* $\frac{7}{15} + \frac{2}{9}$

10 Add *a)* $\frac{2}{5} + \frac{4}{9}$ *b)* $\frac{3}{10} + \frac{5}{18}$ *c)* $\frac{7}{15} + \frac{20}{27}$

4C

1 Subtract *a)* $\frac{1}{2} - \frac{1}{4}$ *b)* $\frac{1}{8} - \frac{1}{10}$ *c)* $\frac{1}{5} - \frac{1}{7}$

2 Subtract *a)* $\frac{1}{2} - \frac{2}{5}$ *b)* $\frac{3}{4} - \frac{2}{3}$ *c)* $\frac{4}{5} - \frac{3}{8}$

3 Subtract *a)* $\frac{1}{8} - \frac{1}{12}$ *b)* $\frac{7}{8} - \frac{5}{12}$ *c)* $\frac{11}{12} - \frac{5}{8}$

4 Subtract *a)* $\frac{1}{6} - \frac{1}{9}$ *b)* $\frac{1}{10} - \frac{1}{15}$ *c)* $\frac{1}{14} - \frac{1}{21}$

5 Subtract *a)* $\frac{2}{6} - \frac{2}{9}$ *b)* $\frac{7}{10} - \frac{7}{15}$ *c)* $\frac{11}{14} - \frac{11}{21}$

6 Subtract *a)* $\frac{1}{8} - \frac{1}{16}$ *b)* $\frac{1}{3} - \frac{1}{9}$ *c)* $\frac{1}{5} - \frac{1}{15}$

7 Subtract *a)* $\frac{3}{8} - \frac{3}{16}$ *b)* $\frac{2}{3} - \frac{2}{9}$ *c)* $\frac{4}{5} - \frac{4}{15}$

8 Subtract *a)* $\frac{5}{12} - \frac{3}{16}$ *b)* $\frac{5}{8} - \frac{3}{20}$ *c)* $\frac{5}{24} - \frac{3}{32}$

9 Subtract *a)* $\frac{2}{3} - \frac{1}{7}$ *b)* $\frac{5}{6} - \frac{3}{14}$ *c)* $\frac{7}{9} - \frac{2}{21}$

10 Subtract *a)* $\frac{1}{2} - \frac{4}{11}$ *b)* $\frac{3}{4} - \frac{7}{22}$ *c)* $\frac{5}{8} - \frac{9}{44}$

4D

1 Write these as mixed numbers in their simplest form:

a) $\frac{3}{2}$ *b)* $\frac{5}{3}$ *c)* $\frac{9}{2}$ *d)* $\frac{10}{4}$ *e)* $\frac{16}{3}$

f) $\frac{9}{7}$ *g)* $\frac{17}{5}$ *h)* $\frac{19}{3}$ *i)* $\frac{15}{6}$ *j)* $\frac{21}{8}$

k) $\frac{29}{5}$ *l)* $\frac{33}{14}$ *m)* $\frac{35}{12}$ *n)* $\frac{41}{8}$ *o)* $\frac{50}{13}$

p) $\frac{64}{24}$ *q)* $\frac{72}{48}$ *r)* $\frac{105}{14}$ *s)* $\frac{121}{22}$ *t)* $\frac{117}{36}$

2 Change these mixed numbers to improper fractions:

a) $1\frac{1}{3}$ *b)* $1\frac{5}{6}$ *c)* $1\frac{3}{4}$ *d)* $2\frac{2}{7}$ *e)* $2\frac{5}{9}$

f) $2\frac{2}{11}$ *g)* $5\frac{1}{5}$ *h)* $4\frac{2}{3}$ *i)* $4\frac{3}{10}$ *j)* $6\frac{2}{7}$

k) $4\frac{9}{11}$ *l)* $5\frac{3}{4}$ *m)* $7\frac{1}{9}$ *n)* $6\frac{3}{4}$ *o)* $6\frac{5}{12}$

p) $9\frac{3}{5}$ *q)* $10\frac{2}{11}$ *r)* $12\frac{4}{7}$ *s)* $15\frac{1}{2}$ *t)* $16\frac{2}{3}$

4E

Find the values of the following in their simplest form:

1 *a)* $\frac{1}{2} \times 2$ *b)* $2 \times \frac{1}{2}$ *c)* $\frac{1}{4} \times 2$ *d)* $2 \times \frac{1}{4}$

2 *a)* $\frac{1}{3} \times 3$ *b)* $3 \times \frac{1}{3}$ *c)* $\frac{2}{3} \times 3$ *d)* $3 \times \frac{2}{3}$

3 *a)* $\frac{1}{5} \times 5$ *b)* $5 \times \frac{1}{5}$ *c)* $\frac{3}{5} \times 5$ *d)* $5 \times \frac{3}{5}$

4 *a)* $\frac{1}{2} \times 6$ *b)* $6 \times \frac{1}{2}$ *c)* $\frac{1}{3} \times 6$ *d)* $6 \times \frac{1}{3}$

5 *a)* $\frac{1}{3} \times 9$ *b)* $9 \times \frac{1}{3}$ *c)* $\frac{2}{3} \times 9$ *d)* $9 \times \frac{2}{3}$

6 *a)* $\frac{1}{5} \times 15$ *b)* $\frac{2}{5} \times 15$ *c)* $\frac{3}{5} \times 15$ *d)* $\frac{4}{5} \times 15$

7 *a)* $\frac{1}{8} \times 24$ *b)* $\frac{3}{8} \times 24$ *c)* $\frac{5}{8} \times 24$ *d)* $\frac{7}{8} \times 24$

8 *a)* $18 \times \frac{1}{6}$ *b)* $18 \times \frac{5}{6}$ *c)* $14 \times \frac{1}{7}$ *d)* $14 \times \frac{3}{7}$

9 *a)* $\frac{1}{2} \times \frac{1}{2}$ *b)* $\frac{1}{2} \times \frac{1}{4}$ *c)* $\frac{1}{3} \times \frac{1}{4}$ *d)* $\frac{1}{3} \times \frac{1}{5}$

10 *a)* $\frac{1}{2} \times \frac{2}{3}$ *b)* $\frac{1}{2} \times \frac{3}{4}$ *c)* $\frac{2}{3} \times \frac{4}{5}$ *d)* $\frac{3}{4} \times \frac{3}{5}$

11 *a)* $\frac{2}{7} \times \frac{3}{4}$ *b)* $\frac{3}{8} \times \frac{4}{7}$ *c)* $\frac{2}{9} \times \frac{3}{8}$ *d)* $\frac{5}{6} \times \frac{3}{10}$

12 *a)* $\frac{2}{3} \times \frac{5}{7}$ *b)* $\frac{5}{8} \times \frac{5}{6}$ *c)* $\frac{5}{9} \times \frac{6}{7}$ *d)* $\frac{8}{9} \times \frac{5}{6}$

13 *a)* $\frac{15}{16} \times \frac{8}{9}$ *b)* $\frac{2}{9} \times \frac{15}{22}$ *c)* $\frac{4}{15} \times \frac{3}{8}$ *d)* $\frac{6}{25} \times \frac{15}{16}$

14 *a)* $\frac{2}{7} \times \frac{4}{5}$ *b)* $\frac{6}{11} \times \frac{3}{22}$ *c)* $\frac{2}{9} \times \frac{6}{14}$ *d)* $\frac{3}{9} \times \frac{5}{7}$

15 *a)* $\frac{6}{8} \times \frac{4}{9}$ *b)* $\frac{3}{5} \times \frac{10}{25}$ *c)* $\frac{8}{9} \times \frac{12}{18}$ *d)* $\frac{9}{12} \times \frac{6}{27}$

4F

Find the value of each of the following:

1 *a)* $1\frac{1}{2} \times \frac{1}{3}$ *b)* $\frac{1}{3} \times 1\frac{1}{2}$ **2** *a)* $1\frac{2}{3} \times \frac{1}{5}$ *b)* $\frac{1}{5} \times 1\frac{2}{3}$

3 *a)* $1\frac{3}{5} \times \frac{1}{8}$ *b)* $\frac{1}{8} \times 1\frac{3}{5}$

4 *a)* $2\frac{1}{4} \times \frac{1}{9}$ *b)* $\frac{1}{9} \times 2\frac{1}{4}$

5 *a)* $3\frac{1}{3} \times \frac{1}{10}$ *b)* $\frac{1}{10} \times 3\frac{1}{3}$

6 *a)* $1\frac{3}{4} \times \frac{1}{7}$ *b)* $1\frac{3}{4} \times \frac{2}{7}$

7 *a)* $2\frac{2}{5} \times \frac{1}{6}$ *b)* $2\frac{2}{5} \times \frac{5}{6}$

8 *a)* $\frac{1}{8} \times 3\frac{5}{9}$ *b)* $\frac{3}{8} \times 3\frac{5}{9}$

9 *a)* $\frac{1}{9} \times 3\frac{3}{8}$ *b)* $\frac{5}{9} \times 3\frac{3}{8}$

10 *a)* $\frac{1}{16} \times 4\frac{4}{5}$ *b)* $\frac{3}{16} \times 4\frac{4}{5}$

4G

Find the value of each of the following in as simple a form as possible:

1 $2\frac{1}{4} \times 2\frac{2}{3}$

2 $3\frac{3}{5} \times 1\frac{1}{9}$

3 $5\frac{1}{5} \times 1\frac{7}{8}$

4 $4\frac{3}{4} \times 1\frac{1}{3}$

5 $2\frac{1}{10} \times 1\frac{2}{3}$

6 $3\frac{1}{9} \times 1\frac{5}{7}$

7 $5\frac{5}{6} \times 3\frac{6}{7}$

8 $1\frac{3}{5} \times 2\frac{1}{4} \times 1\frac{1}{6}$

9 $3\frac{3}{7} \times 1\frac{1}{3} \times 1\frac{5}{16}$

10 $1\frac{7}{15} \times 3\frac{2}{11} \times 1\frac{13}{14}$

4H

Work out the following:

1 *a)* $1 \div \frac{1}{2}$ *b)* $1 \div \frac{1}{3}$ *c)* $1 \div \frac{1}{4}$ *d)* $1 \div \frac{1}{6}$

2 *a)* $2 \div \frac{1}{2}$ *b)* $2 \div \frac{1}{3}$ *c)* $2 \div \frac{1}{4}$ *d)* $2 \div \frac{1}{6}$

3 *a)* $2 \div \frac{2}{3}$ *b)* $3 \div \frac{3}{4}$ *c)* $5 \div \frac{5}{6}$ *d)* $10 \div \frac{5}{6}$

4 *a)* $1 \div \frac{2}{3}$ *b)* $1 \div \frac{3}{4}$ *c)* $1 \div \frac{5}{6}$ *d)* $2 \div \frac{5}{6}$

5 *a)* $4 \div \frac{1}{5}$ *b)* $4 \div \frac{2}{5}$ *c)* $4 \div \frac{4}{5}$ *d)* $2 \div \frac{4}{5}$

6 *a)* $6 \div \frac{2}{3}$ *b)* $3 \div \frac{2}{3}$ *c)* $4 \div \frac{4}{9}$ *d)* $2 \div \frac{4}{9}$

7 *a)* $5 \div \frac{2}{7}$ *b)* $5 \div \frac{1}{7}$ *c)* $3 \div \frac{4}{5}$ *d)* $3 \div \frac{2}{5}$

8 *a)* $8 \div \frac{4}{9}$ *b)* $6 \div \frac{3}{5}$ *c)* $2 \div \frac{4}{5}$ *d)* $3 \div \frac{6}{7}$

9 *a)* $12 \div \frac{8}{11}$ *b)* $12 \div \frac{4}{9}$ *c)* $10 \div \frac{5}{7}$ *d)* $8 \div \frac{4}{7}$

10 *a)* $3 \div \frac{4}{7}$ *b)* $5 \div \frac{2}{7}$ *c)* $6 \div \frac{3}{8}$ *d)* $9 \div \frac{6}{11}$

4I

Write the answer to each of the following in its simplest form:

1 *a)* $3 \div 4\frac{1}{2}$ *b)* $12 \div 4\frac{1}{2}$

2 *a)* $5 \div 2\frac{1}{15}$ *b)* $10 \div 2\frac{1}{15}$

3 *a*) $4 \div 1\frac{1}{8}$ *b*) $2 \div 1\frac{1}{8}$

4 *a*) $9 \div 1\frac{1}{6}$ *b*) $3 \div 1\frac{1}{6}$

5 *a*) $2\frac{1}{3} \div 3\frac{1}{2}$ *b*) $2\frac{1}{3} \div 1\frac{3}{4}$

6 *a*) $1\frac{3}{4} \div 2\frac{5}{8}$ *b*) $3\frac{1}{2} \div 2\frac{5}{8}$

7 *a*) $2\frac{4}{5} \div 2\frac{1}{10}$ *b*) $1\frac{2}{5} \div 4\frac{1}{5}$

8 *a*) $6\frac{3}{4} \div 5\frac{5}{8}$ *b*) $3\frac{3}{8} \div 5\frac{5}{8}$

9 *a*) $1\frac{3}{11} \div 2\frac{7}{22}$ *b*) $2\frac{6}{11} \div 2\frac{7}{22}$

10 *a*) $5\frac{1}{7} \div 6\frac{3}{4}$ *b*) $10\frac{2}{7} \div 6\frac{3}{4}$

4J

Write the answer to each of the following in its simplest form:

1 $4\frac{1}{2} + 3\frac{2}{3}$

2 $2\frac{4}{9} - 1\frac{1}{6}$

3 $5\frac{3}{8} - 2\frac{7}{12}$

4 $2\frac{4}{5} + 4\frac{3}{4} - 3\frac{3}{10}$

5 $6\frac{5}{6} - 1\frac{7}{8} - 1\frac{1}{4}$

6 $2\frac{1}{2} \times 1\frac{1}{5} \div 3\frac{1}{3}$

7 $1\frac{3}{5} \times 2\frac{1}{4} \times 1\frac{7}{8}$

8 $2\frac{2}{9} \times 2\frac{4}{5} \div 1\frac{1}{6}$

9 $(1\frac{1}{4} \div 1\frac{1}{20}) \div 1\frac{7}{18}$

10 $1\frac{5}{7} \times 2\frac{5}{8} \div 1\frac{1}{5}$

* 4K

Find the value of each of the following. Where no brackets are used, multiplication and division take priority over addition and subtraction.

1 $(2\frac{3}{5} + 1\frac{2}{3}) \div 9\frac{3}{5}$

2 $2\frac{2}{3} \times \frac{9}{20} - \frac{3}{10}$

3 $3\frac{3}{4} + 1\frac{7}{8} \div 1\frac{9}{16}$

4 $3\frac{3}{7} \times 1\frac{5}{16} + 1\frac{5}{6}$

5 $1\frac{7}{9} \div 6\frac{2}{3} - \frac{3}{20}$

6 $2\frac{7}{10} \times (1\frac{1}{9} - \frac{5}{6})$

7 $(1\frac{5}{6} + 1\frac{3}{8}) \div 2\frac{1}{16}$

8 $4\frac{2}{5} + 2\frac{2}{5} \times \frac{15}{16}$

9 $(2\frac{1}{4} + 1\frac{5}{6}) \div (4\frac{1}{9} - 3\frac{11}{12})$

10 $(2\frac{1}{6} + 1\frac{1}{15} - 2\frac{3}{10}) \times 2\frac{1}{7}$

4L The Farey Lattice

1 *a*) On a sheet of graph paper, using the first quadrant only, mark the points (2, 1) (3, 1) (3, 2) (4, 1) (4, 2) (4, 3) . . . (7, 1) (7, 2) . . . (7, 6) . . . Continue as far as your scale allows. Note that all these points fall below the 45° line ($y = x$).

b) Let the point (2, 1) represent the fraction $\frac{1}{2}$
" " " (3, 1) " " " $\frac{1}{3}$
" " " (3, 2) " " " $\frac{2}{3}$

What fractions do the following points represent:

(5, 3) (5, 2) (5, 4) (7, 3) (6, 5) (4, 1)?

2 *a*) You can put a straight line from the origin through the points (2, 1) (4, 2) (6, 3) etc. Choose one or two other sets of points which behave similarly. List them and draw the lines.
b) Compare your results with what you found in 4A, numbers 1, 2, 3 etc.

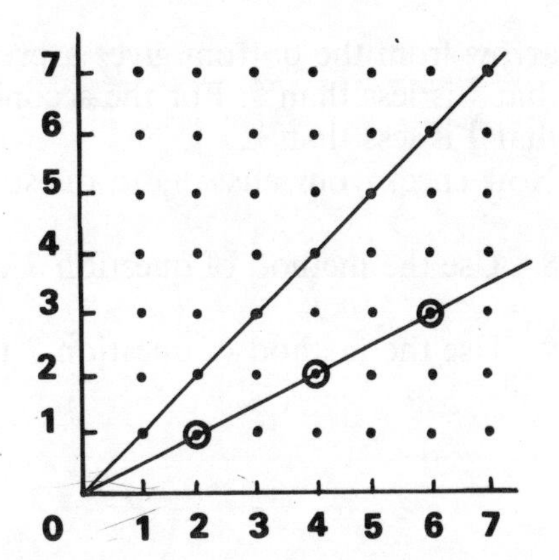

3 *a*) All the points in question 1 fell below the 45° line ($y = x$). If you now marked similar points *above* this line, what could you say about the fractions they represented?
b) What can you say about the points that fall *on* the line $y = x$?
Note The set of points below the line $y = x$ in question 1 is known as a Farey lattice.

4 *a*) Draw lines from the origin to the points (3, 1) and (3, 2). The second line is above the first. This means that the second fraction, $\frac{2}{3}$, is bigger than the first, $\frac{1}{3}$.
b) Draw lines from the origin to the points (4, 3) and (3, 2). Which of the two fractions, $\frac{3}{4}$ and $\frac{2}{3}$, is the bigger?
c) Repeat question *b*) for the fractions $\frac{1}{3}$, $\frac{2}{5}$
d) Repeat question *b*) for the fractions $\frac{3}{7}$, $\frac{1}{2}$
e) Repeat question *b*) for the fractions $\frac{3}{5}$, $\frac{2}{3}$

5 Using the Farey lattice put the following fractions into order of size, smallest first:

a) $\frac{1}{2}$ $\frac{2}{3}$ $\frac{3}{4}$ *b*) $\frac{3}{7}$ $\frac{4}{5}$ $\frac{3}{5}$ *c*) $\frac{1}{4}$ $\frac{1}{5}$ $\frac{2}{7}$

6 You may have to enlarge your Farey lattice for some of the following questions, using a smaller scale and getting in more points.

Arrange in order of size, smallest first:

a) $\frac{3}{4}$ $\frac{5}{6}$ $\frac{2}{3}$ *b*) $\frac{1}{3}$ $\frac{1}{5}$ $\frac{1}{7}$ $\frac{2}{9}$ $\frac{1}{4}$

c) $\frac{3}{8}$ $\frac{1}{3}$ $\frac{2}{5}$ $\frac{3}{7}$ $\frac{4}{9}$ $\frac{4}{7}$ $\frac{1}{2}$ *d*) $\frac{5}{12}$ $\frac{2}{3}$ $\frac{7}{10}$ $\frac{5}{8}$ $\frac{3}{4}$ $\frac{7}{9}$

* 7 The following fractions are in order of size, smallest first:

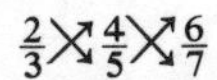

Draw arrows as shown. Multiply together the numbers connected by arrows. The arrow from the top of the first fraction gives a product of 10, while the

arrow from the bottom gives a product of 12. 10 is less than 12. This means that $\frac{2}{3}$ is less than $\frac{4}{5}$. For the second pair the arrows give 28 and 30, showing that $\frac{4}{5}$ is less than $\frac{6}{7}$.
Now check your answers to questions 4*b*), 4*c*) . . . 4*e*) using this method.

∗ *8* Use the method of question 7 to check your answers to question 5.

∗ *9* Use the method of question 7 to check your answers to question 6.

4M

1 Anna spent half of her pocket money on Saturday, and half of the remainder on Monday. What fraction was left to last the rest of the week?

2 Tony had a bag of sweets. He gave away one third of the sweets, ate one third of those that were left and then lost the rest.
What fraction of the full bag did he lose? If there were 18 sweets in the bag to begin with, how many did he eat himself?

3 Of the cars in a car park, $\frac{1}{12}$ were blue, $\frac{1}{12}$ were red, $\frac{1}{18}$ were black, $\frac{1}{6}$ were white, $\frac{1}{18}$ were grey, $\frac{1}{9}$ were brown and $\frac{2}{9}$ were yellow. If all the rest were green, what fraction of the total were green?
From this information, what can you deduce about the total number of cars in the car park, if it cannot hold more than 80 cars?

4 A motorist had completed $\frac{2}{3}$ of a journey when her car broke down. She managed to get a lift for half the remainder of the journey, but then had to walk the rest of the way. What fraction of the journey did she have to walk?
If her lift was for $7\frac{1}{2}$ km, how far did she walk and how long was the whole journey?

5 A greengrocer received a delivery of tomatoes on a Thursday during a heat wave. That day he sold $\frac{1}{3}$ of these, on Friday he sold $\frac{1}{4}$ of the original delivery, and on Saturday half as many as he sold on Thursday. His refrigerator was out of action, and by Monday the rest of the tomatoes had gone bad. What fraction of the original consignment did he throw away?
If he started with 60 kg, how many did he sell on Saturday?

6 Four girls shared some prize money. Liz had half the total amount, Ann had half as much as Liz, and Carol and Diane, who had tied for third place, shared what was left equally between them. What fraction of the whole amount did Carol receive?

7 $\frac{2}{9}$ of the chocolates in Mary's box had hard centres and so did $\frac{3}{7}$ of those in Jane's box. If Mary's box had 36 chocolates and Jane's had 42, how many soft-centred chocolates did each girl have?
What fraction of the total of all the chocolates had hard centres?

8 $\frac{1}{3}$ of the pupils in a class had dark hair and $\frac{9}{10}$ of these had brown eyes. $\frac{1}{2}$ of the other pupils in the class also had brown eyes. What fraction of the whole class had brown eyes?

If there are 9 pupils with both brown eyes and dark hair, how many are there altogether in the class?

9 $\frac{3}{5}$ of the tiles on a kitchen floor are white, $\frac{1}{4}$ of the remainder are blue and the rest red. What fraction of the floor is red?

If there are 150 tiles altogether, how many are blue?

10 Half the pieces of chalk in a box were white. $\frac{1}{3}$ of the coloured ones were blue, $\frac{1}{4}$ of the rest were red, and the others were yellow.

If there were 6 yellow ones, how many pieces of chalk were there altogether?

5 Decimals and Metric Measure

5A Decimals

1 Multiply 2·5 by *a)* 10 *b)* 100 *c)* 1000 *d)* 0·1 *e)* 0·01

2 Multiply 23·4 by *a)* 10 *b)* 100 *c)* 0·1 *d)* 0·01 *e)* 0·001

3 Multiply 0·3 by *a)* 0·01 *b)* 0·1 *c)* 10 *d)* 100 *e)* 1000

4 Multiply 0·42 by *a)* 0·1 *b)* 10 *c)* 100 *d)* 1000 *e)* 10 000

5 Multiply 1·53 by *a)* 10 *b)* 100 *c)* 1000 *d)* 0·1 *e)* 0·01

6 Multiply 0·06 by *a)* 10 *b)* 100 *c)* 1000 *d)* 10 000 *e)* 0·1

7 Multiply 1·04 by *a)* 0·01 *b)* 0·1 *c)* 10 *d)* 100 *e)* 1000

8 Multiply 42·3 by *a)* 10 *b)* 100 *c)* 0·1 *d)* 0·01 *e)* 0·001

9 Multiply 627 by *a)* 10 *b)* 100 *c)* 0·1 *d)* 0·01 *e)* 0·001

10 Multiply 0·0072 by *a)* 10 *b)* 100 *c)* 1000 *d)* 10 000 *e)* 0·1

5B

1 Multiply 1·2 by *a)* 2 *b)* 20 *c)* 200 *d)* 0·2 *e)* 0·02

2 Multiply 21·3 by *a)* 3 *b)* 30 *c)* 0·3 *d)* 0·03 *e)* 0·003

3 Multiply 0·5 by *a)* 5 *b)* 50 *c)* 500 *d)* 0·5 *e)* 0·05

4 Multiply 2·24 by *a)* 2 *b)* 20 *c)* 200 *d)* 0·2 *e)* 0·02

5 Multiply 37·6 by *a)* 4 *b)* 40 *c)* 0·4 *d)* 0·04 *e)* 0·004

6 Multiply 0·03 by *a)* 3 *b)* 30 *c)* 300 *d)* 3000 *e)* 0·3

7 Multiply 3·82 by *a)* 2 *b)* 20 *c)* 200 *d)* 0·2 *e)* 0·02

8 Multiply 27·4 by *a)* 5 *b)* 50 *c)* 0·5 *d)* 0·05 *e)* 0·005

9 Multiply 0·72 by *a)* 3 *b)* 30 *c)* 300 *d)* 0·3 *e)* 0·03

10 Multiply 0·085 by *a)* 7 *b)* 70 *c)* 700 *d)* 7000 *e)* 0·7

5C

1 Multiply 1·23 by *i)* 2 *ii)* 3

Using these two answers, write down the answers to the following:

a) $1{\cdot}23 \times 23$ *b)* $1{\cdot}23 \times 230$ *c)* $1{\cdot}23 \times 32$
d) $1{\cdot}23 \times 3{\cdot}2$ *e)* $1{\cdot}23 \times 0{\cdot}23$

2 Multiply 2·4 by *i)* 5 *ii)* 2

Hence write down the answers to:

a) $2{\cdot}4 \times 5{\cdot}2$ *b)* $2{\cdot}4 \times 52$ *c)* $2{\cdot}4 \times 0{\cdot}52$
d) $2{\cdot}4 \times 2{\cdot}5$ *e)* $2{\cdot}4 \times 250$

3 Multiply 36·2 by *i)* 5 *ii)* 3

Hence find the answers to the following:

a) $36{\cdot}2 \times 53$ *b)* $36{\cdot}2 \times 530$ *c)* $36{\cdot}2 \times 3{\cdot}5$
d) $36{\cdot}2 \times 0{\cdot}53$ *e)* $36{\cdot}2 \times 30{\cdot}5$

4 Multiply 71·6 by *i)* 3 *ii)* 8

Hence find the answers to the following:

a) $71{\cdot}6 \times 38$ *b)* $71{\cdot}6 \times 8{\cdot}3$ *c)* $71{\cdot}6 \times 8{\cdot}03$
d) $71{\cdot}6 \times 30{\cdot}8$ *e)* $71{\cdot}6 \times 830$

5 Multiply 9·78 by *i)* 7 *ii)* 4 *iii)* 3

Hence find the answers to the following:

a) $9{\cdot}78 \times 7{\cdot}4$ *b)* $9{\cdot}78 \times 740$ *c)* $9{\cdot}78 \times 43$
d) $9{\cdot}78 \times 0{\cdot}43$ *e)* $9{\cdot}78 \times 34{\cdot}7$

6 Multiply 201·4 by *i)* 8 *ii)* 5

Hence find the answers to the following:

a) $201{\cdot}4 \times 8{\cdot}5$ *b)* $201{\cdot}4 \times 85$ *c)* $201{\cdot}4 \times 58$
d) $201{\cdot}4 \times 0{\cdot}58$ *e)* $201{\cdot}4 \times 5{\cdot}08$

7 Multiply 32·9 by *i)* 7 *ii)* 6

Hence find the value of

a) $32{\cdot}9 \times 7{\cdot}6$ *b)* $32{\cdot}9 \times 0{\cdot}76$ *c)* $32{\cdot}9 \times 6{\cdot}07$
d) $32{\cdot}9 \times 60{\cdot}7$ *e)* $32{\cdot}9 \times 66$

8 Multiply 4·72 by *i)* 2 *ii)* 7

Hence find the value of

a) $4{\cdot}72 \times 27$ *b)* $4{\cdot}72 \times 2{\cdot}7$ *c)* $4{\cdot}72 \times 7{\cdot}2$
d) $4{\cdot}72 \times 0{\cdot}207$ *e)* $4{\cdot}72 \times 0{\cdot}722$

9 Multiply 6·09 by *i*) 3 *ii*) 5 *iii*) 6

Hence find the value of

a) $6{\cdot}09 \times 3{\cdot}5$ *b*) $6{\cdot}09 \times 56$ *c*) $6{\cdot}09 \times 0{\cdot}36$
d) $6{\cdot}09 \times 36{\cdot}5$ *e*) $6{\cdot}09 \times 5{\cdot}5$

10 Multiply 87·3 by *i*) 2 *ii*) 5 *iii*) 8

Hence find the value of

a) $87{\cdot}3 \times 2{\cdot}8$ *b*) $87{\cdot}3 \times 8{\cdot}02$ *c*) $87{\cdot}3 \times 0{\cdot}052$
d) $87{\cdot}3 \times 258$ *e*) $87{\cdot}3 \times 0{\cdot}505$

5D

Work out the value of the following:

1 $2{\cdot}92 \times 3{\cdot}4$
2 $46{\cdot}7 \times 0{\cdot}29$
3 $24{\cdot}1 \times 6{\cdot}8$
4 $29{\cdot}8 \times 0{\cdot}071$
5 $301{\cdot}5 \times 0{\cdot}57$
6 $6{\cdot}23 \times 4{\cdot}9$
7 $7{\cdot}83 \times 0{\cdot}62$
8 $0{\cdot}38 \times 0{\cdot}56$
9 $1{\cdot}009 \times 2{\cdot}88$
10 $13{\cdot}6 \times 2{\cdot}07$

5E

1 Divide 29·8 by *a*) 10 *b*) 100 *c*) 0·1 *d*) 0·01
2 Divide 363·7 by *a*) 10 *b*) 100 *c*) 1000 *d*) 0·01
3 Divide 2342 by *a*) 10 *b*) 100 *c*) 1000 *d*) 10000
4 Divide 9·46 by *a*) 10 *b*) 100 *c*) 0·1 *d*) 0·01
5 Divide 59·07 by *a*) 10 *b*) 100 *c*) 1000 *d*) 0·1
6 Divide 5206 by *a*) 10 *b*) 100 *c*) 1000 *d*) 10000
7 Divide 2·08 by *a*) 10 *b*) 100 *c*) 0·1 *d*) 0·01
8 Divide 0·63 by *a*) 10 *b*) 0·1 *c*) 0·01 *d*) 0·001
9 Divide 0·028 by *a*) 10 *b*) 100 *c*) 0·1 *d*) 0·01
10 Divide 5·634 by *a*) 10 *b*) 100 *c*) 1000 *d*) 0·1

5F

1 Divide 36·9 by *a*) 3 *b*) 30 *c*) 0·3 *d*) 0·03
2 Divide 155 by *a*) 5 *b*) 50 *c*) 500 *d*) 0·5
3 Divide 2·52 by *a*) 3 *b*) 30 *c*) 0·3 *d*) 0·03
4 Divide 2240 by *a*) 7 *b*) 70 *c*) 700 *d*) 0·7
5 Divide 5·22 by *a*) 9 *b*) 0·9 *c*) 0·09 *d*) 0·009

6 Divide 0·444 by *a*) 12 *b*) 1·2 *c*) 0·12 *d*) 0·012

7 Divide 7·42 by *a*) 7 *b*) 70 *c*) 0·7 *d*) 0·07

8 Divide 7·42 by *a*) 14 *b*) 140 *c*) 1·4 *d*) 0·14

9 Divide 57·6 by *a*) 8 *b*) 80 *c*) 800 *d*) 0·8

10 Divide 57·6 by *a*) 16 *b*) 160 *c*) 1600 *d*) 1·6

5G

Find the value of the following:

1 19·8 ÷ 11

2 19·8 ÷ 2·2

3 84·8 ÷ 1·6

4 10·75 ÷ 0·25

5 6·12 ÷ 1·7

6 50·7 ÷ 0·39

7 2·652 ÷ 0·13

8 3·206 ÷ 0·07

9 3·206 ÷ 0·14

10 0·868 ÷ 1·4

5H

Work out the answers to the following. (If the last figure is 0 it must be included.)

1 3·1 × 6·2

2 2·53 × 7·9

3 4·09 × 0·73

4 6·913 × 1·5

5 0·017 × 1·23

6 29·007 × 3·028

7 1·5 × 2·2 × 2·08

8 Write down the number of decimal places in each of the answers you have obtained.

9 What is the rule for finding the number of decimal places in any multiplication sum?

10 Without working the following out in full, write down the number of decimal places there should be in each answer:

a) 2·9 × 3·45 *b*) 6·29 × 1·052
c) 1·52 × 6·73 *d*) 2·095 × 12·63
e) 25·063 × 0·058 *f*) 29·356 × 1·05
g) 3·6 × 4·7 × 5·3 *h*) 17·8 × 396·2
i) 16·3 × 8·9 × 14·0 *j*) 0·006 × 13·7 × 5·6

5I

In this exercise 5's are to be 'rounded upwards' where necessary.

1 Correct these numbers to 1 decimal place:

26·64 295·77 6·08 7·98 405·62
0·35 10·29 0·96 25·08 162·35

2 Correct these numbers to 2 decimal places:

2·863	5·091	27·267	321·098	0·005
1·004	0·083	13·666	4·999	5·108

Give the answers to the following correct to 2 decimal places:

3 $67·23 \div 14$ ***4*** $4·96 \div 2·3$ ***5*** $0·542 \div 3·9$

6 $0·631 \div 7·8$ *7* $7·39 \div 8·1$ ***8*** $0·056 \div 0·23$

9 $3·17 \div 13$ ***10*** $0·0611 \div 0·94$

5J

1 Write the following numbers correct to 2 significant figures:

42·3	178	0·177	224	1345
6·09	0·0136	21·9	627	24 362

2 Write the following numbers correct to 3 significant figures:

421·4	3·729	0·6422	3427	16229
4999	1·298	0·7021	14550	5687

Write down the answers to the following correct to 2 significant figures:

3 $11·2 \times 5·1$ **4** $170 \times 2·2$ **5** $5·8 \times 9·7$

6 225×19 7 $492 \div 2·6$ **8** $4·3 \div 1·8$

9 $19·3 \div 6·1$ ***10*** $0·75 \div 1·4$

5K Fractions to Decimals

1 Write the following fractions as decimals:

a) $\frac{1}{10}$ *b)* $\frac{1}{100}$ *c)* $\frac{1}{1000}$ *d)* $\frac{3}{10}$ *e)* $\frac{5}{100}$

f) $\frac{17}{100}$ *g)* $\frac{9}{1000}$ *h)* $\frac{129}{1000}$ *i)* $\frac{2377}{10\,000}$ *j)* $\frac{39}{1000}$

2 Write the following fractions as decimals:

a) $\frac{1}{2}$ *b)* $\frac{1}{4}$ *c)* $\frac{1}{8}$ *d)* $\frac{1}{16}$ *e)* $\frac{1}{32}$ *f)* $\frac{1}{64}$ *g)* $\frac{3}{4}$

h) $\frac{5}{8}$ *i)* $\frac{7}{8}$ *j)* $\frac{9}{16}$ *k)* $\frac{3}{16}$ *l)* $\frac{11}{32}$ *m)* $\frac{17}{64}$

3 Write the following fractions as decimals:

a) $\frac{1}{5}$ *b)* $\frac{1}{50}$ *c)* $\frac{1}{500}$ *d)* $\frac{1}{25}$ *e)* $\frac{1}{250}$ *f)* $\frac{1}{125}$

g) $\frac{3}{5}$ *h)* $\frac{4}{5}$ *i)* $\frac{7}{25}$ *j)* $\frac{16}{25}$ *k)* $\frac{47}{125}$ *l)* $\frac{13}{50}$

m) $\frac{29}{50}$ *n)* $\frac{6}{125}$ *o)* $\frac{19}{125}$ *p)* $\frac{217}{500}$

4 Change the following decimals to fractions, cancelling down to the lowest terms:

a) 0·5 *b*) 0·25 *c*) 0·75 *d*) 0·125 *e*) 0·375 *f*) 0·875

g) 0·05 *h*) 0·15 *i*) 0·45 *j*) 0·85 *k*) 0·325 *l*) 0·825

m) 0·555 *n*) 0·3125 *o*) 0·9375

5 *a*) Add the following fractions: $\frac{1}{4}$ $\frac{3}{8}$ $\frac{1}{2}$ $\frac{7}{8}$

b) Convert each of the above fractions to a decimal and add the decimals together.
c) Write your answer to *a*) as a decimal.
Does it agree with the answer to *b*)?

6 Repeat question 5 using the following fractions:

$\frac{2}{5}$ $\frac{7}{25}$ $\frac{17}{125}$ $\frac{3}{10}$.

7 Repeat question 5 using these fractions: $\frac{3}{4}$ $\frac{3}{8}$ $\frac{2}{5}$ $\frac{1}{10}$

8 Change to decimals: $\frac{5}{16}$ $\frac{7}{32}$ $\frac{8}{125}$ $\frac{19}{25}$ $\frac{7}{40}$ $\frac{23}{80}$ $\frac{19}{20}$

9 Change to fractions: 0·128 0·355 0·328 0·555 0·096 0·0072

10 All the above fractions give 'terminating' decimals. Try and turn $\frac{1}{3}$ into a decimal and you will find the decimal never terminates. Try also $\frac{1}{7}$, $\frac{1}{11}$ and $\frac{1}{13}$. These decimals are called 'recurring decimals' and they are studied in Book 2.

You will note that the denominators of all fractions which give terminating decimals are multiples of 2 or 5 only. Check this with all the fractions used in questions 1 to 9 above.

5L Metric Units

1 Write each of these lengths in metres:

a) 1 m 3 cm *b*) 2 m 5 cm 3 mm *c*) 6 cm 6 mm *d*) 46 cm
e) 15 mm *f*) 725 mm *g*) 2 km *h*) 5 km 2 m
i) 7 km 5 m *j*) 12 km 4 m 3 cm

2 Write these weights in grams:

a) 2 kg *b*) 2 kg 5 g *c*) 12·6 kg *d*) 0·375 kg
e) 0·017 kg *f*) 0·573 kg *g*) 2·81 kg *h*) 4 kg 816 g
i) 1 kg 13 g *j*) 2 kg 4 g

3 Write these lengths in centimetres:

a) 2 m 5 cm *b*) 42 m 6 cm *c*) 6 cm 4 mm *d*) 36 mm
e) 18 cm 2 mm *f*) 5 m 4 mm *g*) 0·004 m *h*) 1·2 mm
i) 7 m 16 cm *j*) 300 m

4 Write these weights in kilograms:

a) 550 g *b*) 75 g *c*) 2 kg 40 g *d*) 12·6 g
e) 5·8 g *f*) 1 g *g*) 2380 g *h*) 17000 g
i) 916 g *j*) 217 g

5 How many square centimetres are there in a square metre?

6 How many square metres are there in a square kilometre?

7 How many cubic centimetres are there in:

a) 2·5 litres *b*) 7·36 litres *c*) 100 litres
d) $\frac{1}{4}$ litre *e*) 0·375 litre?

8 Write these liquid quantities in litres:

a) 1200 cm^3 *b*) 400 cm^3 *c*) 750 cm^3
d) 4·520 cm^3 *e*) 70 cm^3

9 A model is made for a hydro-electric scheme which is being planned, the scale being 1:10,000. What actual lengths are represented by:

a) 5 cm *b*) 15 cm *c*) 3·5 cm?

10 A map is drawn on the scale 1:100 000. What distances on the map represent:

a) 2 km *b*) 80 km *c*) 300 km?

11 How many pieces of tape each 13 cm long can be cut from a two metre length, and how much tape will be left over?

12 Add the following:

a) 3 km 45 m; 4 km 517 m; 8 km 803 m
b) 1 m 9 cm 4 mm; 3 m 23 cm 2 mm; 25 mm
c) 25 mm; 77 mm; 43 mm; 11 mm (Give your answer in cm.)
d) 2 kg 480 g; 3 kg 17 g; 15 kg 800 g
e) 5 litres 48 ml; 17 litres 36 ml; 816 ml

13 Carry out the following multiplications:

a) 427 cm × 18 (Give your answer in metres.)
b) 36 m 18 cm by 350 (Give your answer in km to 2 d.p.)
c) 28 mm by 240 (Give your answer in metres to 2 d.p.)
d) 346 g by 27 (Give your answer in kg to 2 d.p.)
e) 43·2 kg by 170 (Give your answer in tonnes to 2 d.p.)

14 Carry out the following subtractions:

a) 429 g from 2 kg *b*) 3 kg 5 g from 5 kg *c*) 295 m from 1·5 km
d) 85 cm from 2·4 m *e*) 365 cm^3 from half a litre

15 *a*) What is $\frac{1}{5}$ of 2 litres? Give your answer in ml.
b) What is $\frac{2}{3}$ of 3·6 km? Answer in metres.
c) What is $\frac{3}{4}$ of 1 tonne 840 kg? Answer in kg.
d) Divide 5 m 10 cm by 17. Answer in cm.
e) What is $\frac{4}{7}$ of a kilogram? Answer in grams to the nearest gram.

16 A camper needs 17 pieces of nylon cord, each 43 cm long. What is the total length of cord he needs? If the cost is 4p a metre and the length he buys must be in whole metres, how much will the cord cost?

17 A student wishes to cover the floor of her room with cork tiles each 30 cm square. If the room measures 4·2 metres square, how many tiles does she need for the floor altogether?

18 Repeat question 17 for a room 5·1 metres square. Draw a sketch to illustrate your answer.

19 A group of boys call at a café to buy drinks. They can buy Coke at 5p a glass, or large bottles of Coke for 75p, but they must then use their own cups. If the glasses hold 200 ml and the large bottle holds five litres, which is the best buy? Could any other factors beside cost affect their decision as to which to buy?

20 A teacher moves into a farm cottage surrounded by fields. He wishes to put a strand of barbed wire on the boundary fence right round his garden. The boundary is just under 400 m in length. He wants to know the weight of the barbed wire to see whether he can fetch it from the nearby town in his Mini. So he weighs a piece of wire 75 cm long and finds it weighs 84 g. The wire is wound on a reel weighing 2 kg and each reel contains 200 m of wire.

If his Mini is designed to carry four adults, can he safely fetch the wire? Can he take his wife with him?

6 Angles, Bearings, Triangles, Polygons

6A Angles

1 Using your protractor, draw these angles in your book:

a) 25° *b*) 125° *c*) 50° *d*) 100°
e) 150° *f*) 250° *g*) 80° *h*) 280°

Label each one 'acute', 'obtuse' or 'reflex'.

2 Using a ruler and pencil only, draw these angles as accurately as you can:

a) a quarter of a revolution
b) a third of a revolution
c) a sixth of a revolution
d) two thirds of a revolution
e) a twelfth of a revolution

Measure accurately the angles you have drawn and write down by how many degrees your angle varies from the correct one in each case.

3 Using a ruler and pencil only, draw these angles as accurately as you can:

a) 20° *b*) 60° *c*) 110° *d*) 145° *e*) 230° *f*) 320°

Measure each angle accurately and write down by how many degrees your angle is incorrect.
Label each one 'acute', 'obtuse' or 'reflex'.

4 Draw for yourself four irregular pentagons (5 sides). Carefully measure the angles in each and find the total. You should find that the four totals are approximately the same. If they are not, check your measurements. What do you think that the total should be?

5 Repeat question 4, this time drawing four seven-sided figures.

6 Write down the smallest angle between the hands of a clock at

a) 3 o'clock *b*) 8 o'clock *c*) 1.10
d) 3.40 *e*) 6.20 *f*) 5.55

7 Write down the angle turned through by the minute hand of a clock in

a) 10 min *b*) 25 min *c*) 45 min
d) 55 min *e*) 65 min *f*) 90 min

8 The speedometer in a car is graduated from 0 to 160 km/h in 280°.

a) What angle has the pointer turned through from rest to a speed of

i) 20 km/h *ii*) 32 km/h *iii*) 80 km/h *iv*) 100 km/h?

b) As the car accelerates from 24 km/h the pointer on the speedometer turns through 70°. What speed has then been reached?

9 *a*) The dial on bathroom scales turns through one complete revolution when a man weighing 120 kg steps on it. What angle does the dial turn through to record these weights:

i) 42 kg *ii*) 58 kg *iii*) 75 kg?

b) When the dial turns through these angles, what weights are on the scales:

i) half a revolution *ii*) two thirds of a revolution
iii) 240° *iv*) $1\frac{1}{4}$ revolutions

6B Bearings

1 Draw accurately in your books a diagram to show the 8 main compass points: north, south, east, west, north-east, north-west, south-east and south-west.

2 Using the diagram of question 1, write down the bearings corresponding to:

a) 135° clockwise from N. *b*) 90° anticlockwise from NE.
c) 225° clockwise from S. *d*) 45° clockwise from E.
e) 180° anticlockwise from SW. *f*) 270° anticlockwise from SW.
g) 315° clockwise from W.

3 Some of the parts of question 2 would have been simpler if you had not been given reflex angles. Rewrite these parts in a simpler way.

4 Draw accurate diagrams to show these bearings:

a) N 35° E. *b*) N 85° E. *c*) S 25° W.
d) N 10° W. *e*) S 65° W. *f*) S 45° E.

5 Write each of the bearings in question 4 as 'three-figure bearings'.

6 Draw accurate diagrams to show these bearings:

a) 015° *b*) 335° *c*) 260° *d*) 095° *e*) 170° *f*) 220°

7 *B* is a coastguard station 8 km due east of another coastguard station *A*. At a certain time a ship is sighted from *A* on a bearing of 040° and from *B* on a bearing of 310°. Find how far the ship is from *A*.
Use a scale of 1 cm to 1 km.

8 To find his exact position, the captain of a ship took two bearings. He found the bearing of a lighthouse was 060°, and the bearing of a lightship was 130°. From his charts he knew that the lighthouse was 10 km due north of the lightship. How far from each was his ship?

9 From two points R and S on the south coast it was possible to see a ship anchored at sea. The bearing of the ship from R was 110° and from S it was 165°. If S is 6 km due east of R, how far was the ship from R and from S?

10 To find his exact position on a map, a hiker took the bearings of two distant landmarks. The bearing of a church spire was 340° and that of a river bridge was 260°. From his map he knew that the church spire was 16 km due NE of the bridge. How far was he from the bridge?

11 A ship sailed from a port A for 14 km on a bearing of 030° to reach an island B. From B it sailed another 18 km on a bearing of 140° to an island C. How far was C from A and on what course should the ship sail to reach A directly after leaving C? Use a scale of 1 cm to 2 km.

12 A hiker followed a stream on a bearing of 110° for 2 km and came to the remains of a Roman road which was on a bearing of 190°. This he followed for 2·5 km. He left this road and followed another track on a bearing of 240° for 1·5 km. What was the shortest distance back to the starting point and on what bearing would he have to walk to get there? Use a scale of 2 cm to 1 km.

13 An aeroplane's next refuelling stop was 1600 km away on a bearing of 315°, when it had to alter course to avoid storms. It travelled for 900 km on a bearing of 335°. How far had it still to go to its refuelling stop and on what bearing did it have to fly?

14 A plane flew from an airport A on a bearing of 225° for 3000 km to reach airport B. It then flew 2400 km on a bearing of 160° to reach airport C. If it had not had to call at B, on what bearing should it have been flown and over what distance, in order to land at C?

15 A ship sailed from a port P to a port Q 20 km away on a bearing of 235°. After leaving Q it sailed another 35 km on a bearing of 080° to reach port R. What was the distance and bearing of R from P?

16 Using a scale of 1 cm to 50 km on both axes, plot the following points:

A (0, 0) B (300, 200) C (800, 0) D (1100, 500) E (600, 700)

Imagine that a ship follows the route from A to B, B to C, C to D, D to E and E to A, and give the bearings and distances for each part. Take the y axis to be due north.

17 Repeat question 16 using the points:

A (400, 400) B (300, 0) C (700, 300) D (400, 1000) E (0, 600).

18 Repeat question 16 using the points:

A (600, 600) B (100, 400) C (100, 800) D (800, 200) E (100, 0).

6C Triangles, Supplementary Angles

1 Draw two acute-angled and two obtuse-angled triangles. In each, measure the angles as accurately as you can and find their sum. If you had measured the angles exactly, what do you think this sum should be?

Note In questions 2 onwards the figures are not drawn accurately so angles must be *calculated*, not measured.

2 Calculate the values of $a, b, c, d \ldots j$.
Describe each triangle, e.g. right angled, scalene, etc.

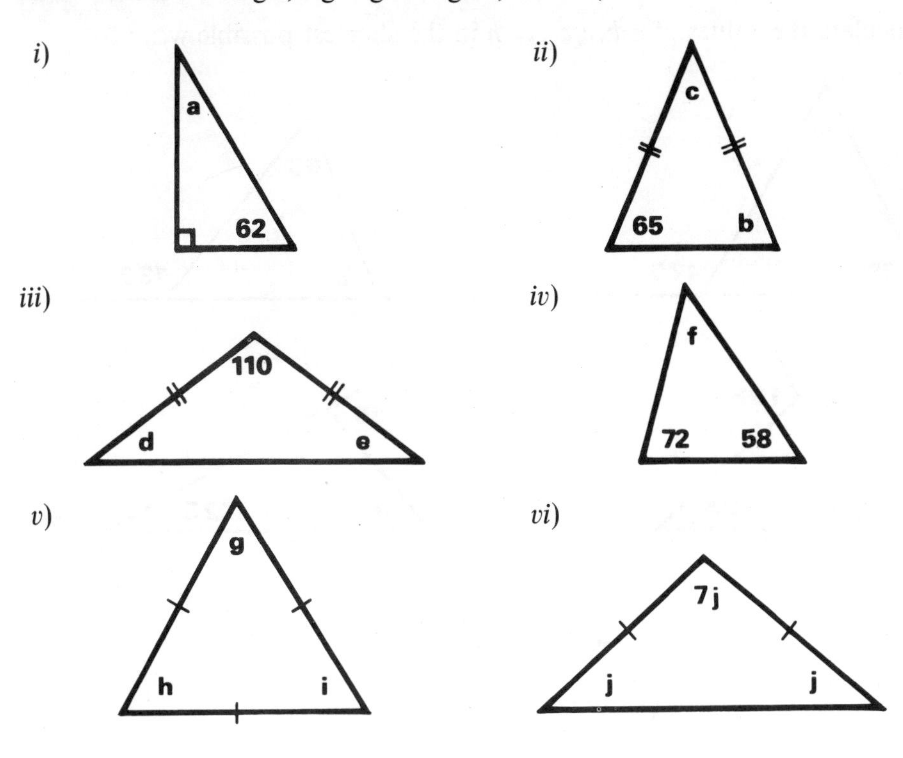

3 Calculate the values of a, b, c and d.

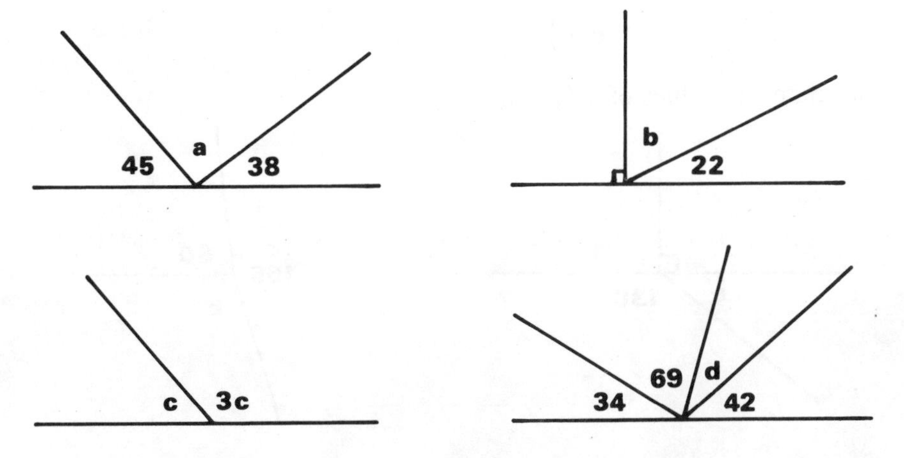

4 Using the same diagram each time find the values of c and d in each of these cases:

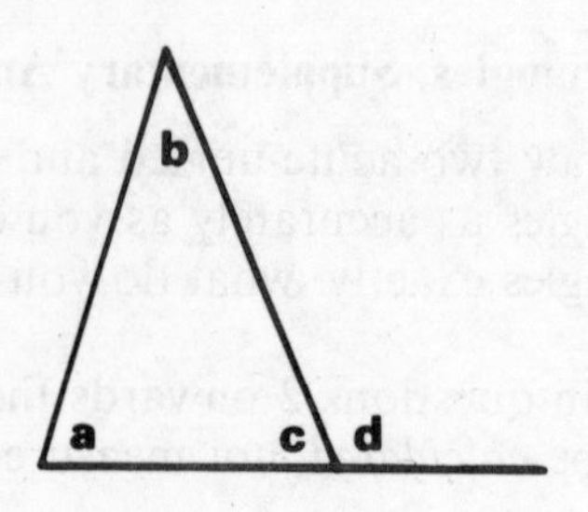

i) $a = 40°$; $b = 70°$
ii) $a = 50°$; $b = 65°$

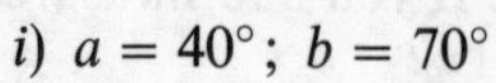

iii) $a = 75°$; $b = 30°$
iv) $a = 38°$; $b = 42°$
v) $a = 25°$; $b = 110°$

What do you notice about the value of d and the sum $a+b$ in each case?

5 Calculate the values of $a, b, c, d \ldots h$ in the shortest possible way:

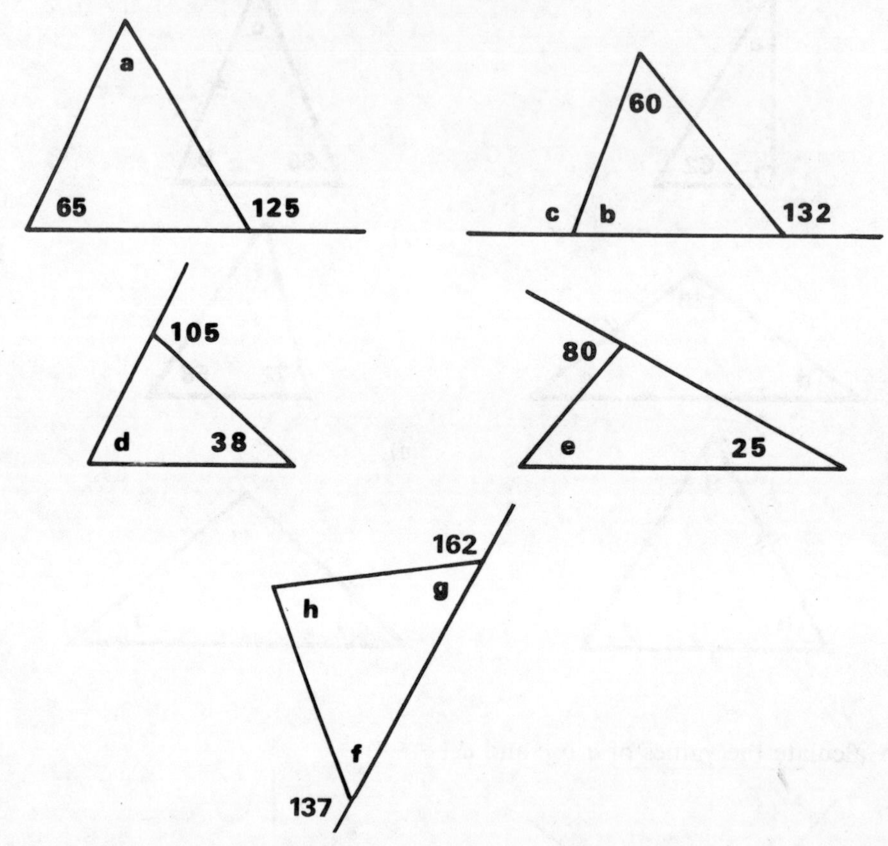

6 Calculate the values of $a, b, c, \ldots i$.

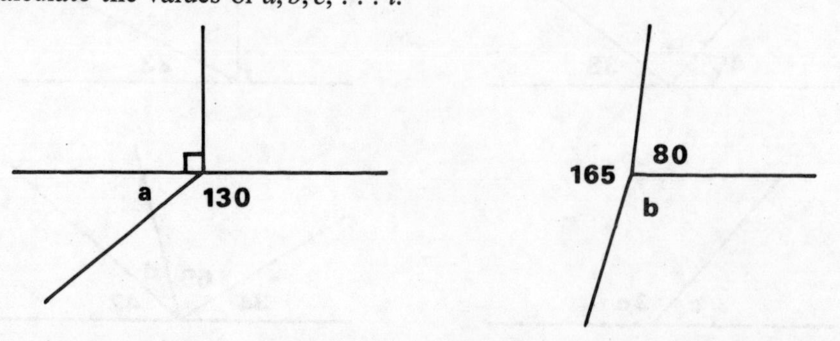

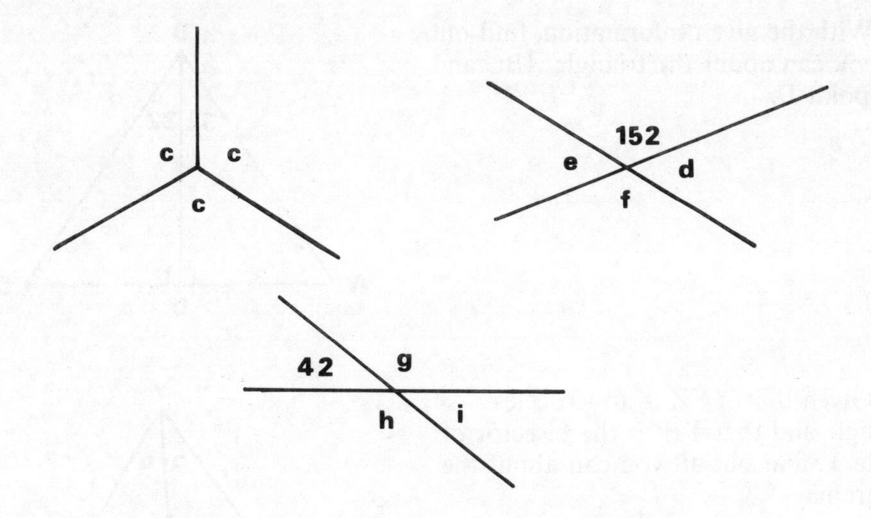

7 Calculate the values of all the angles denoted by letters:

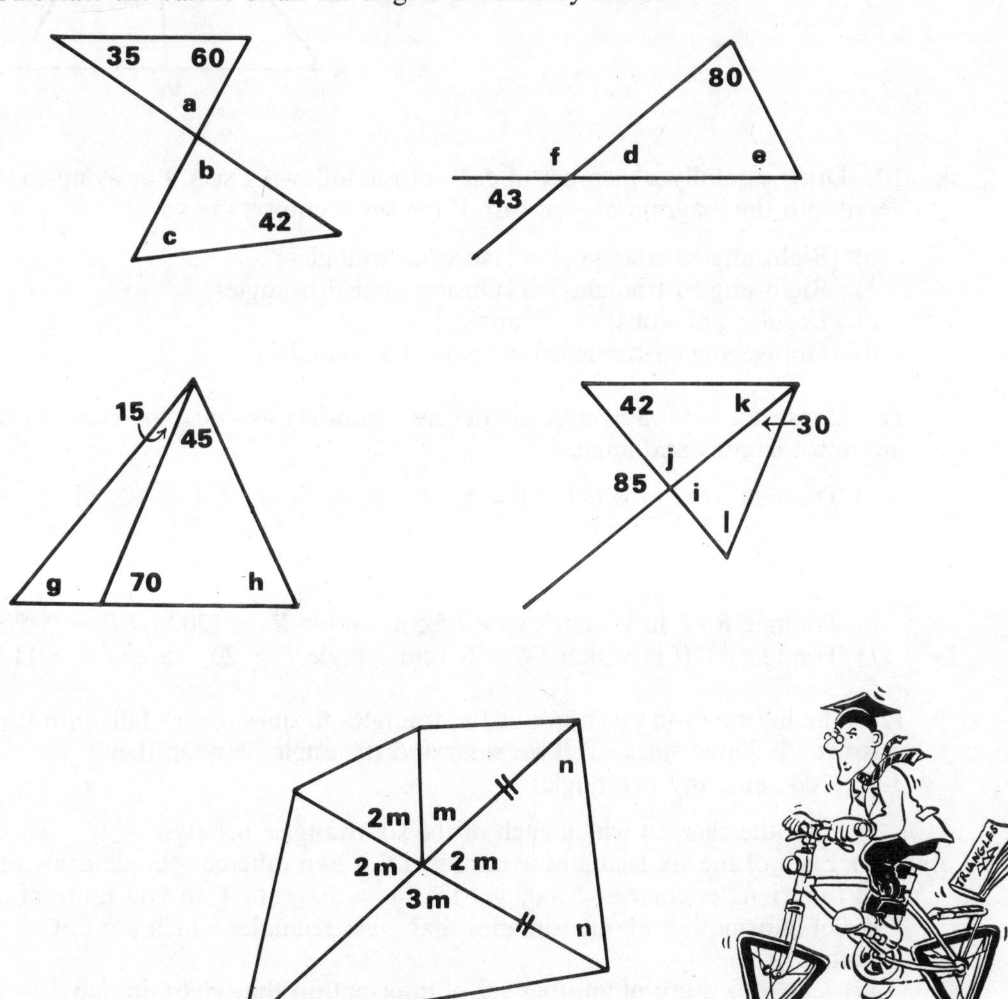

8 With the given information, find out all you can about the triangle ABC and the point D.

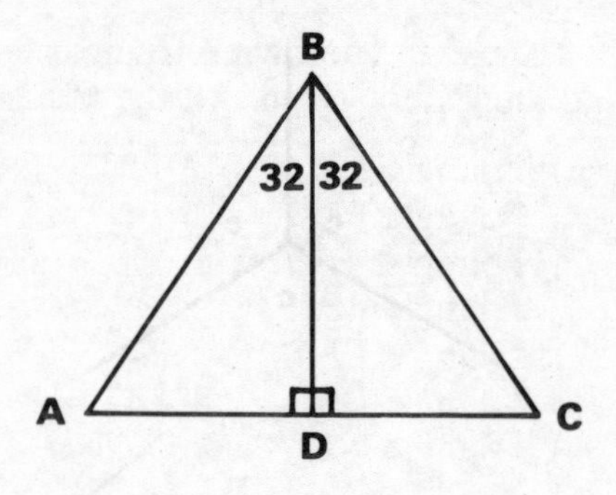

9 Given that XYZ is an isosceles triangle and that YW is the bisector of angle Y, find out all you can about the diagram.

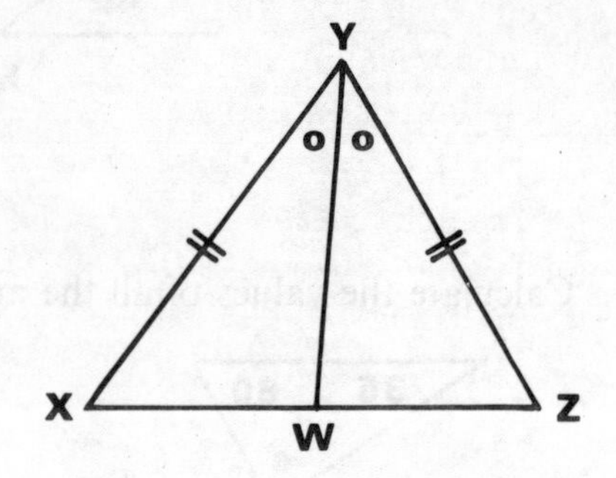

* *10* Draw carefully a member of each of the following sets. Put as much detail into the diagram as you can. If the set is empty, say so:

a) {Right angled triangles} ∩ {Isosceles triangles}
b) {Right angled triangles} ∩ {Obtuse angled triangles}
c) {Regular polygons} ∩ {Triangles}
d) {Obtuse angled triangles} ∩ {Isosceles triangles}

11 Construct the following triangles as accurately as you can. Measure the unknown lengths and angles.

a) Triangle ABC in which $AB = 6$ cm $BC = 5$ cm $CA = 4{\cdot}5$ cm
b) Triangle LMN in which $LM = 3$ cm $MN = 5$ cm $NL = 6{\cdot}2$ cm
c) Triangle PQR in which $PQ = 5{\cdot}5$ cm angle $P = 48°$ angle $Q = 52°$
d) Triangle XYZ in which $XY = 5{\cdot}8$ cm angle $Y = 70°$ $YZ = 4{\cdot}9$ cm
e) Triangle RST in which $RS = 4{\cdot}5$ cm angle $R = 120°$ $TR = 5{\cdot}5$ cm
f) Triangle FGH in which $FG = 6{\cdot}3$ cm angle $F = 20°$ angle $G = 115°$

** *12* The information given about the triangles in question 11 falls into three classes: 1 Three sides 2 Two sides and the angle between them 3 One side and any two angles.

i) State the class to which each of the six triangles belongs.
ii) Each of the six triangles is unique, i.e. if two different people draw one of the triangles correctly, they get the same diagram. Can you think of a set of information about triangles that gives triangles which are not unique?
iii) Can you think of another set of information that gives unique triangles?

13 Draw the following triangles accurately to scale. Measure the lengths and angles not given. All the triangles are unique.

a) Triangle LMN with $LM = 7{\cdot}2$ cm $MN = 6{\cdot}5$ cm angle $M = 40°$
b) Triangle BCD with $BC = 5{\cdot}8$ cm angle $B = 48°$ angle $D = 75°$
c) Triangle AEH with $AE = 6{\cdot}1$ cm $EH = 7{\cdot}3$ cm $HA = 6$ cm
d) Triangle PST with $PS = 4{\cdot}2$ cm $ST = 7{\cdot}5$ cm angle $P = 110°$
e) Triangle ELM with $EL = 5{\cdot}3$ cm $LM = 5{\cdot}3$ cm angle $L = 64°$
f) Triangle DOG with $DO = 7{\cdot}7$ cm angle $D = 51°$ angle $O = 71°$
g) Triangle TOP with $TO = 6{\cdot}7$ cm angle $O = 130°$ $TP = 11{\cdot}4$ cm

** *14* The following triangles either cannot be drawn at all or, if drawn, are not unique. If they can be drawn, draw them and state how many solutions are possible. If they cannot be drawn, say why.

a) Triangle EFG with angle $E = 60°$ angle $F = 38°$ angle $G = 82°$
b) Triangle DEH with $DE = 8{\cdot}2$ cm $EH = 4{\cdot}8$ cm $HD = 3{\cdot}2$ cm
c) Triangle PIG with $PI = 4{\cdot}7$ cm $IG = 5{\cdot}9$ cm angle $G = 38°$
d) As triangle *c*) but with angle $G = 60°$
e) Triangle SPG with angle $S = 70°$ angle $P = 81°$ angle $G = 40°$

15 A parallelogram has sides of 7 cm and 5 cm and an angle of 68°. Draw the parallelogram accurately to scale and measure the two diagonals.

16 The pentagon $PQRST$ has sides PQ, QR, RS, ST, TP of lengths 3, 4, 5, 4, 3 cm respectively.

a) Is this sufficient information to draw the pentagon?
b) Given that $PR = PS = 5$ cm, draw the pentagon and measure the five angles.
c) Given that angle $Q = 100°$ and angle $R = 90°$, draw the pentagon and measure the other three angles and the lengths of the diagonals through P.

17 Draw the quadrilateral $KLMN$ in which KL, LM, MN and NK are of lengths 3·4, 4·7, 5·1 and 4·3 cm respectively, and diagonal KM is 6·2 cm. Measure the four angles and the other diagonal.

18 Repeat question 17 making angle $L = 90°$, and the four sides as stated. Measure the two diagonals and the other angles.

6D Parallel Lines

1 Draw this diagram accurately in your book. Join AY and measure the acute angles you have formed at A and Y. Are they equal?

If the given lengths had been unequal, would the angles at A and Y still have been equal?

AB and XY are parallel. Would the angles at A and Y still be equal if AB and XY were not parallel?

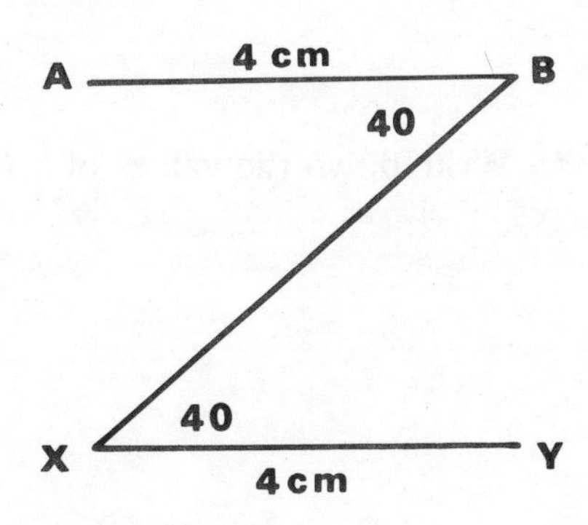

2 Calculate the values of the angles $a, b, c, \ldots k$.

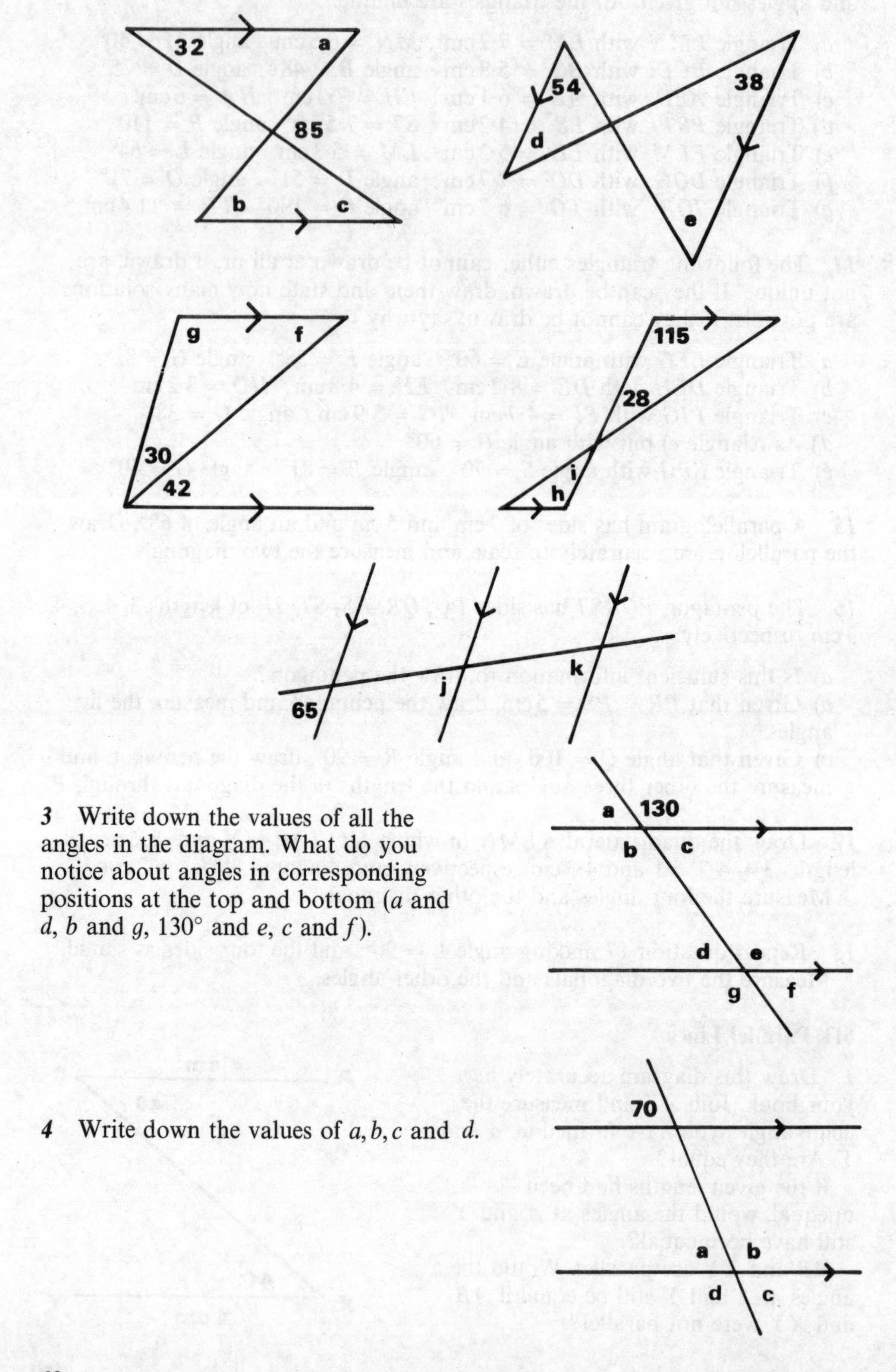

3 Write down the values of all the angles in the diagram. What do you notice about angles in corresponding positions at the top and bottom (a and d, b and g, 130° and e, c and f).

4 Write down the values of a, b, c and d.

5 By finding pairs of alternate angles and pairs of corresponding angles, write down the values of *a*, *b*, *c*, *d* and *e*. What do you notice about the opposite angles in the parallelogram?

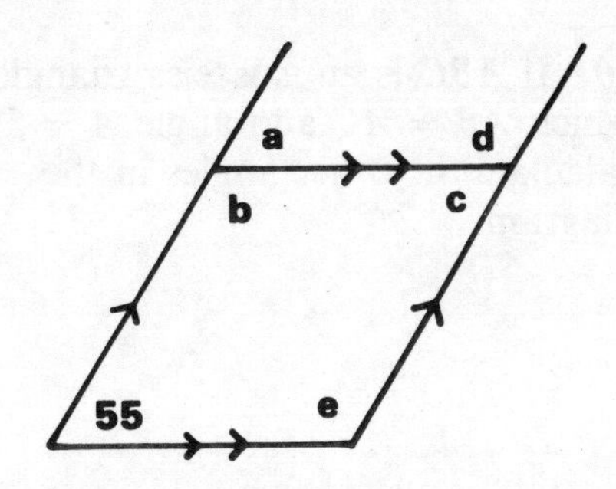

6 Write down the values of *a*, *b*, *c*, *d* and *e*.

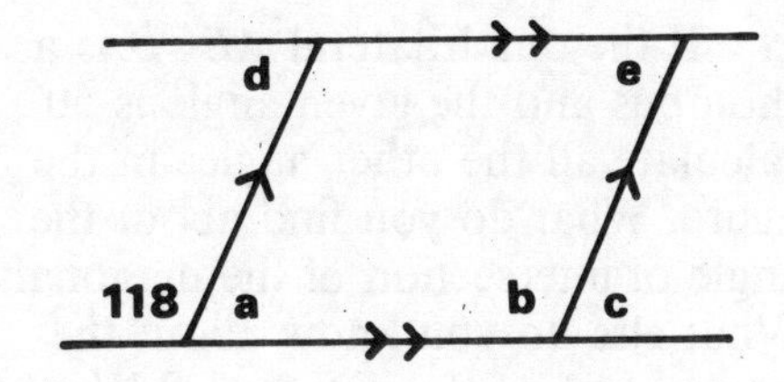

7 Write down pairs of equal angles in this diagram. State also any other relationships you know between these angles.

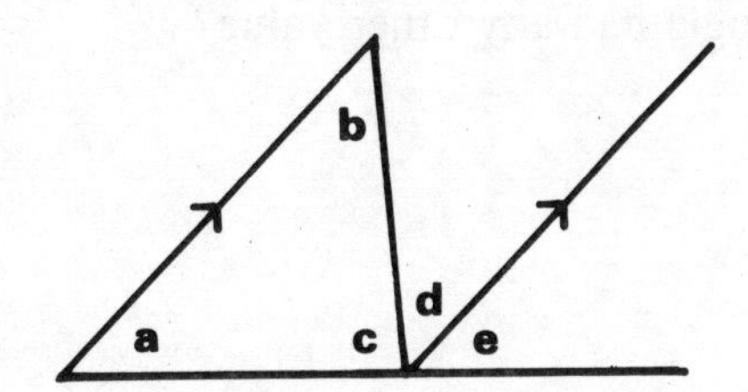

8 If *ABC* is an isosceles triangle in which *AB* = *AC*, write down the values of the other angles in the diagram.

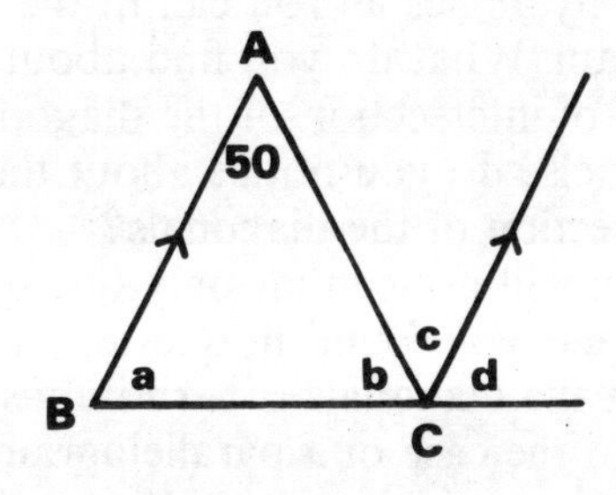

9 If *ABCD* is a rhombus, i.e. all the sides are of equal length, write down the angles of the rhombus.

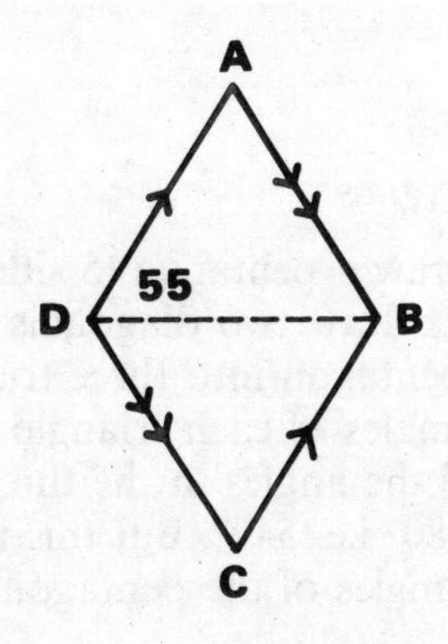

10 If ABC is an isosceles triangle in which $AB = AC$ and angle $A = 58°$, calculate the other angles in the diagram.

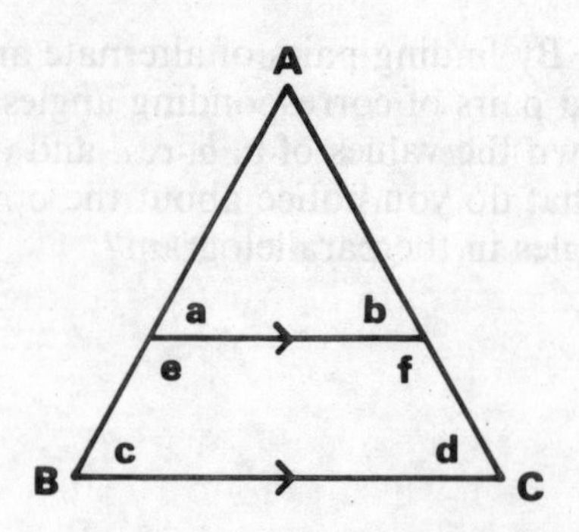

11 If the quadrilateral $ABCD$ is a rhombus and the given angle is 50°, calculate all the other angles in the figure. What do you find about the angle of intersection of the diagonals? What else do you know about the intersection of the diagonals? Would these two things still be true if the given angle had any other value?

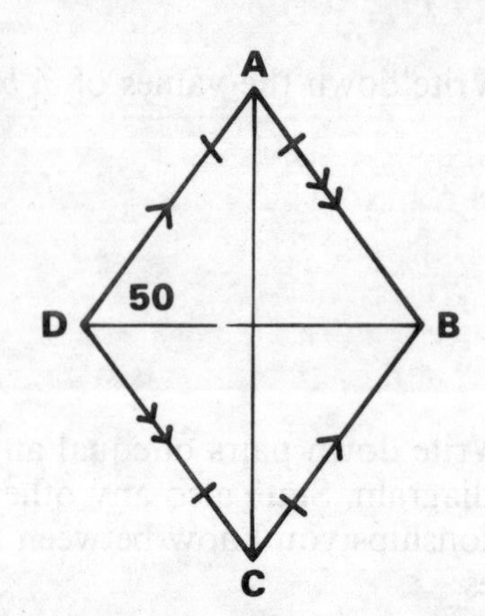

12 If $PQRS$ is a parallelogram and the angles shown are 40° and 30°, calculate as many angles as you can in the diagram. What do you find about the angle of intersection of the diagonals? What else do you notice about the intersection of the diagonals?

You will notice that only one of the two facts you found in question 11 about the diagonals of a rhombus is true in the case of a parallelogram which is not a rhombus. Why is this?

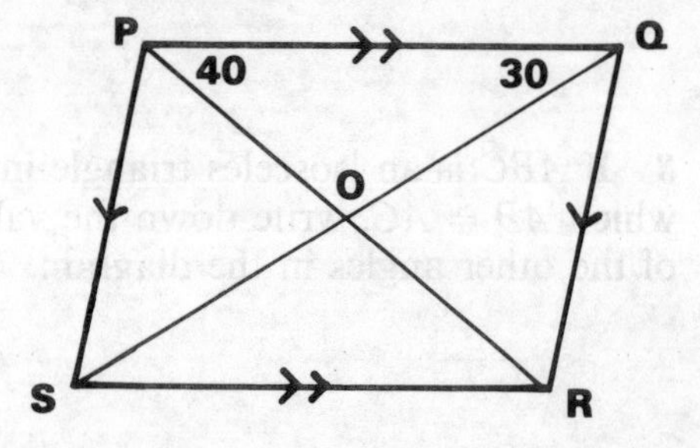

6E Polygons

1 *a*) Draw a pentagon (5 sides). From one vertex draw two diagonals as shown to divide the pentagon into three triangles. The sum of the angles of each triangle is 180°. So the sum of all the angles in the three triangles is $3 \times 180°$ i.e. 540°. But this is the sum of the five angles of the pentagon.

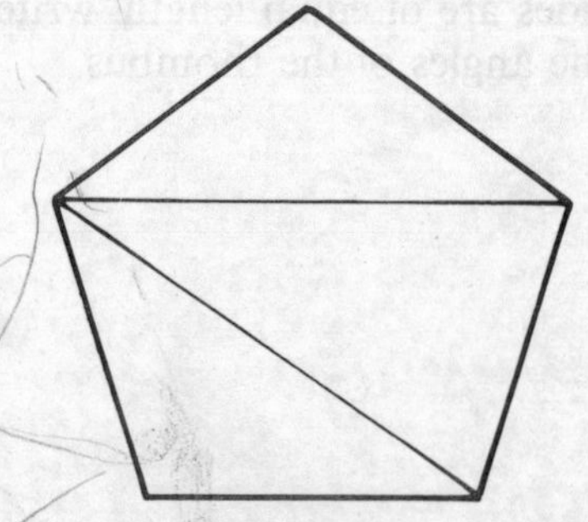

b) In a similar way find the sum of all the angles of a hexagon (6 sides) and an octagon (8 sides).
c) If each of these figures had been regular, all the sides would have been equal and all the angles would have been equal. So each angle of a regular pentagon would be 540° divided by 5, i.e. 108°. Using the answers you calculated in *b*), what is the angle of a regular hexagon and a regular octagon?

2 Draw a circle and mark its centre. Inside the circle draw by eye a pentagon which is approximately regular. Join the five vertices to the centre to make five triangles.

a) If the pentagon were truly regular, what would be the value of each of the five angles at the centre?
b) The five triangles are isosceles. Why?
c) From *a*) what is the 'angle at the vertex' for each of these five isosceles triangles?
d) What is the base angle of each triangle? Hence what is the size of each angle of the pentagon? Your answer should agree with the answer found in question 1*a*).

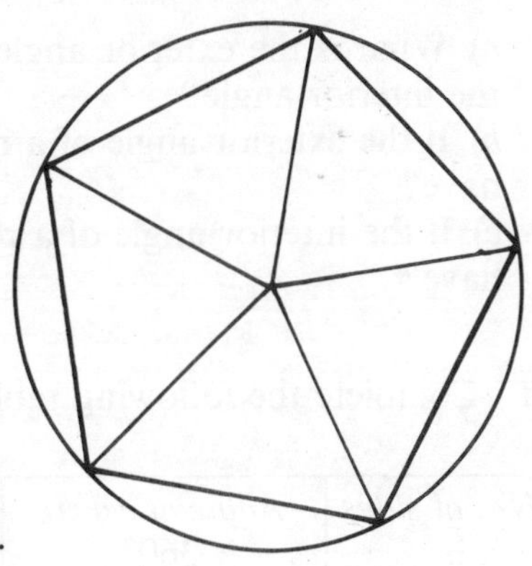

3 Repeat question 2 starting with an octagon.

4 Repeat question 2 starting with a hexagon. What is special about the triangles in a regular hexagon?

5 We can now see how to draw regular polygons quite simply. Draw a circle. If the polygon has *n* sides, draw *n* radii at equal angles. Thus for an octagon these would be spaced at $\frac{360°}{8}$ i.e. 45° intervals.
Join the tips of these radii to form the polygon.

a) Using this method, draw circles of 3 cm radius and construct accurately a regular pentagon and a regular octagon.
b) There is a special method for constructing a regular hexagon. Do you know what it is?

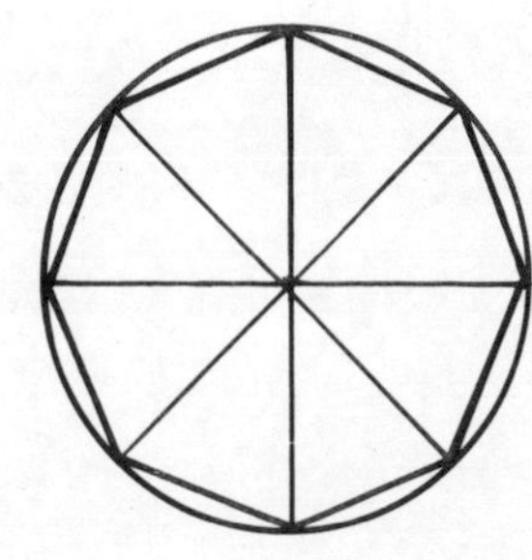

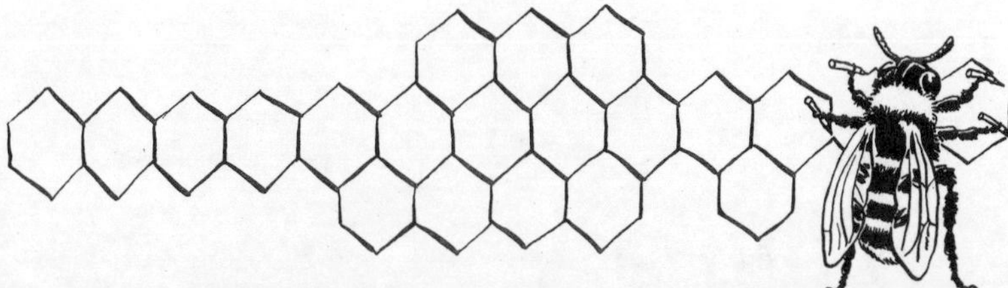

6 An exterior angle of a polygon is formed by producing one of the sides as shown in the diagram. What are the exterior angles of a regular pentagon, hexagon and octagon? Compare the size of these exterior angles with the angles at the centre you found in questions 2, 3 and 4. They should be the same.

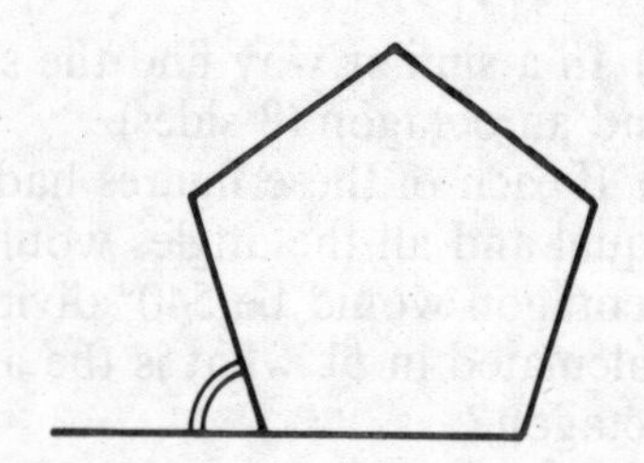

7 We now have a quick method of finding both the exterior angle and the interior angle of any regular polygon of n sides.

1 Divide 360° by n to get the angle at the centre.
2 This is also the exterior angle.
3 Subtract from 180° to get the interior angle.

a) What is the exterior angle of a regular polygon with 12 sides? What is the interior angle?
b) If the exterior angle of a regular polygon is 40°, how many sides does it have?
c) If the interior angle of a regular polygon is 144°, how many sides does it have?

8 Complete the following table for n = 3, 4, 5, 6, 8, 9, 10.

No. of sides n	*Angle at centre* $\frac{360°}{n}$	*Exterior angle*	*Interior angle*	*Name of polygon*
3				
4				
5	72°	72°	108°	Regular pentagon
6				
etc.				

Miscellaneous Examples A

A1

1 State which of these pairs of fractions has the larger value:

a) $\frac{3}{4}\ \frac{4}{5}$ *c)* $\frac{5}{8}\ \frac{13}{20}$ *e)* $\frac{2}{7}\ \frac{4}{13}$
b) $\frac{5}{6}\ \frac{7}{9}$ *d)* $\frac{7}{11}\ \frac{19}{30}$

2 The following calculations appear in a book of arithmetic from the planet Nova. The first one is worked out correctly. Find out what base the Novenians use and complete the other calculations.

$$\begin{array}{r}15\\+22\\\hline 40\\\hline\end{array}\qquad\begin{array}{r}33\\+14\\\hline \\\hline\end{array}\qquad\begin{array}{r}23\\-15\\\hline \\\hline\end{array}\qquad\begin{array}{r}132\\-\ 33\\\hline \\\hline\end{array}\qquad\begin{array}{r}36\\\times\ 2\\\hline \\\hline\end{array}$$

3 Find the angle each letter represents. Give brief explanations of your calculations.

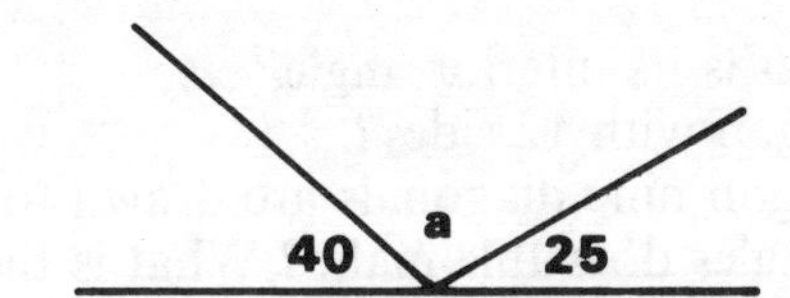

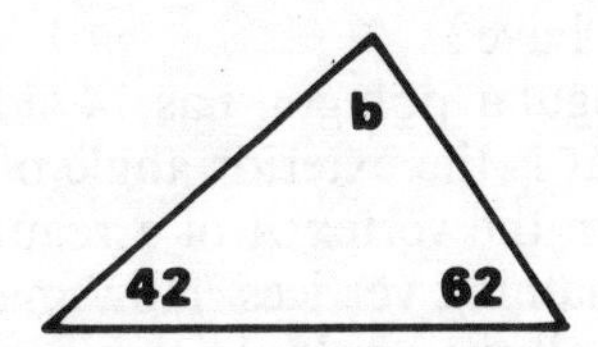

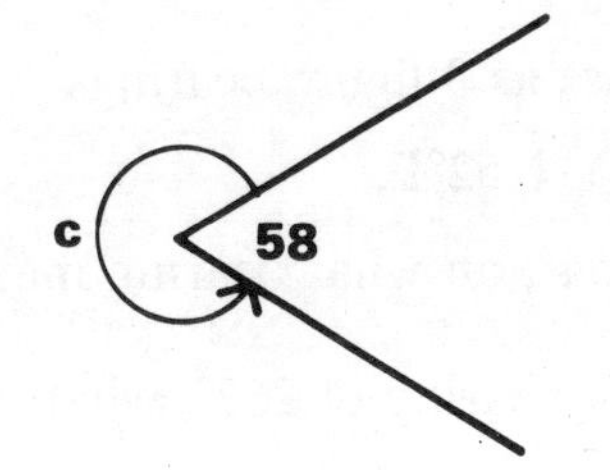

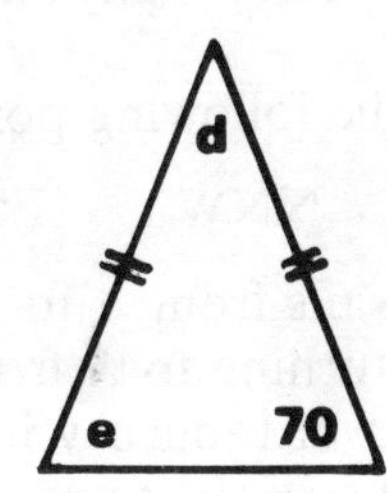

4 P is the set of prime numbers less than 30.
Q is the set of odd numbers less than 30.
R is the set of multiples of 5 less than 30.
S is the set of square numbers less than 30.

a) List the members of the sets P, Q, R, S.
b) List the members of each of the sets stated:

i) $P \cap Q$ *ii)* $P \cap R$ *iii)* $Q \cap S$ *iv)* $P \cap S$ *v)* $R \cap S$

c) Give the value of *i)* $n(P)$ *ii)* $n(S)$.
d) State whether the following statements are TRUE or NOT TRUE:

i) $17 \in P$ *ii)* $20 \notin S$ *iii)* $P \subset Q$

5 Thin wire is being wound on a large drum. At each turn 1 m 15 cm of wire is wound on.

a) How much wire is wound on in *i*) 100 turns *ii*) 800 turns *iii*) 900 turns?
b) How many turns will it take to wind on one kilometre of wire? Give your answer to the nearest turn above.
c) Why was a large drum and thin wire specified?

A2

1 Express the following as decimals of a kilometre:

a) 50 m *b*) 400 m *c*) 400 cm *d*) 2 m *e*) 1000 mm

Express the following as decimals of a kilogram:

f) 358 g *g*) 10 g *h*) 500 mg *i*) 900 g *j*) $28\frac{1}{2}$ g

2 *a*) The interior angle of a regular polygon is 135°. How many sides does it have?
b) A regular polygon has 14 sides. What is its interior angle?
c) What is the exterior angle of a polygon with 12 sides?
d) From the vertex *A* of a regular polygon nine diagonals are drawn to the remaining vertices. How many triangles does this make? What is the sum of all the angles in these triangles? How many vertices are there in the polygon? How many sides? What is the interior angle?

3 *a*) Express the following points of the compass as 3 figure bearings:

NW. NNW. ESE. S38°W. N52°E.

b) If a ship sails from *A* to *B* on a bearing 154°, on what bearing must it sail when returning to *A* from *B*?
c) If the outward journey in *b*) had been on a bearing of 262°, what would be the bearing of the return journey?

* ***4*** The house system is introduced into a school, and it is possible for the pupils to win house points for good work in the classroom, at games, or for other activities. There are four houses, with colours orange, red, pale green and blue. House points are recorded each week on a special board. For each house a small pennant of appropriate colour is inserted for every house point won. Five small pennants are replaced by one larger 'second size' pennant, five of these by one still larger 'third size', and five of these by a still larger 'fourth size'. Not more than five of any size are allowed. What is the maximum number of points that can be recorded in this manner?

Interest grows and the number of house points recorded weekly increases. It is decided to allow six pennants of each size before they are replaced by a larger one. How many points can now be recorded altogether for each house?

What parts of the above information are superfluous?

5 If $A = \{$One legged objects in your home$\}$
$B = \{$Two legged objects in your home$\}$
$C = \{$Three legged objects in your home$\}$
$D = \{$Four legged objects in your home$\}$
$E = \{$Living creatures in your home$\}$

write down not more than three members of each set. Some sets may be empty.

Describe in words the sets $A \cap E$, $B \cap E$, $C \cap E$, $D \cap E$.
Which of these four sets would normally be ϕ?
If $A \cap E$ is not ϕ what does it mean?

A3

1 A bakery makes twenty kinds of cakes. Of these, five contain both butter and eggs but not fruit, four contain eggs and fruit but not butter, three contain butter and fruit but not eggs, one contains fruit but no butter or eggs, and none contain butter or eggs only without the other two.

Draw a Venn diagram to represent these facts, and from it deduce how many kinds contain all three of these ingredients.

2 Perform the following calculations in the bases stated:

a) Base 4 | Base 7 | Base 2 | Base 8

$$\begin{array}{r} 32 \\ +23 \\ \hline \\ \hline \end{array} \qquad \begin{array}{r} 56 \\ -21 \\ \hline \\ \hline \end{array} \qquad \begin{array}{r} 1101 \\ \times 1011 \\ \hline \\ \hline \end{array} \qquad \begin{array}{r} 37 \\ \times 45 \\ \hline \\ \hline \end{array}$$

Give a denary check for each.

b) State the base in which these calculations are carried out, and provide the missing figure:

$$\begin{array}{r} 34 \\ +\ 2* \\ \hline 111 \\ \hline \end{array} \qquad \begin{array}{r} 110 \\ -\ 5* \\ \hline 11 \\ \hline \end{array} \qquad \begin{array}{r} 30 \\ \times *2 \\ \hline 2220 \\ \hline \end{array}$$

3 A girl receives a box of sweets for her birthday. She gives one third of them to her younger brother, one sixth to a friend at school, one to each of three teachers and two each to her mother and father. She herself eats one third of the sweets.

How many did the box contain? Was she generous?

4 A girl buys 130 cm of blue ribbon, 250 cm of red ribbon and enough green ribbon to make the total length up to 5 metres. If the cost is 75p altogether and all three ribbons cost the same per metre, what did the green ribbon cost? If the ribbon was sold in minimum lengths of half a metre, what would the green ribbon cost?

5 Five lightships are moored in a dangerous stretch of sea. Their code names are Herring, Cod, Plaice, Mackerel and Haddock. The bearing of Cod from Herring is 046°, of Plaice from Cod is 162°, of Mackerel from Plaice is 232°, and of Haddock from Mackerel is 307°.

a) It is required to find the bearing of Herring from Haddock. Is this possible? If not, what further information do you need?
b) If you are told that the distances from Herring to Cod, Cod to Plaice, Plaice to Mackerel and Mackerel to Haddock are all equal, can you now find the bearing of Herring from Haddock?

A4

1 If $A = \{\text{Letters in a word}\}$ $B = \{\text{Vowels in the same word (excluding y)}\}$ and $C = (\text{Consonants in the same word (including y)}\}$ give examples of words for which the following are true:

a) $n(A \cap C) = 0$ *b*) $n(A \cap B) = n(A \cap C) = 1$ *c*) $A \cap B = A$
d) $n(A \cap B) = 2 \times n(A \cap C)$ *e*) $n(A \cap B) = 2,\ n(A \cap C) = 6$
f) $A \cup B \cup C = A$ *g*) $n(A \cap C) = 2 \times n(A \cap B)$

2 Find the angles marked with a letter in the figures below. Give brief reasons for your answers:

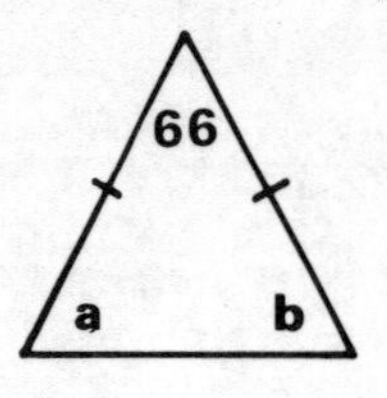

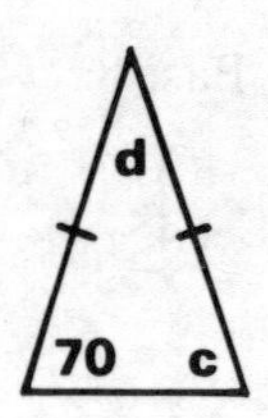

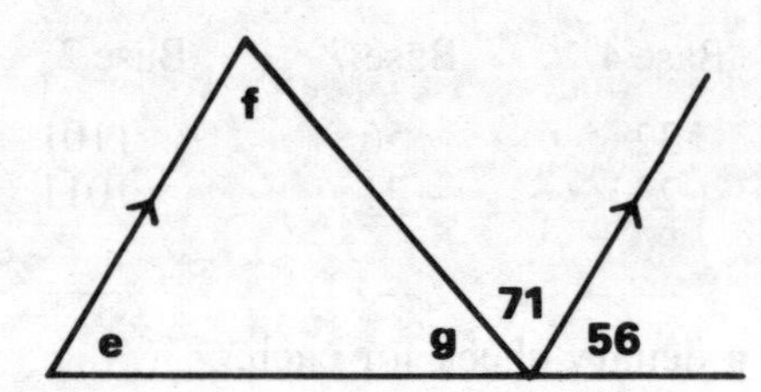

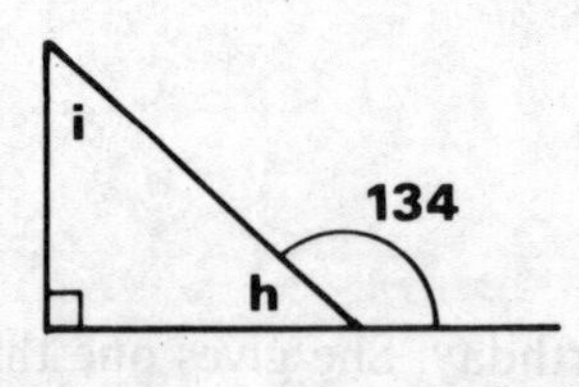

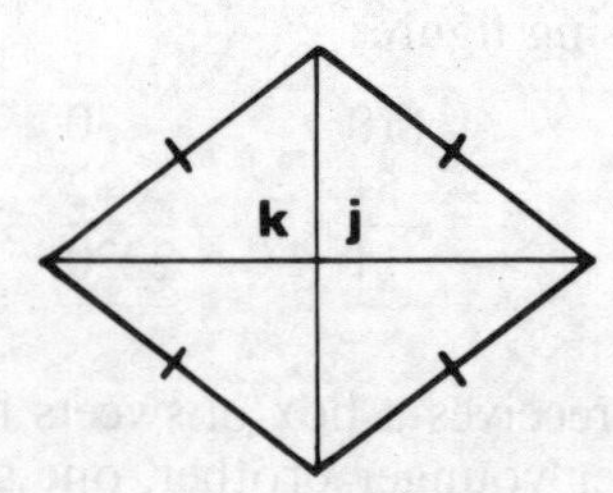

3 Express the following decimals as fractions:

a) 0·375 *b*) 0·75 *c*) 0·67 *d*) 0·6 *e*) 0·85

Find the value of the following:

f) 0·37 of £2 *g*) 0·55 of 3 metres *h*) 0·75 of 2 kg 200 g
i) 0·1 of 85p *j*) 0·85 of an hour

4 A race of 2000 metres starts from a point a little distance round the first lap and ends at the end of the last lap. When the winner is breasting the tape at the end of the fourth lap, one competitor is still only three quarters of the way round this lap, and another is two thirds of the way round.

a) If the distance between them is $42\frac{1}{2}$ metres, what is the length of the lap?
b) How far is the starting point round the first lap?

5 The number 121 in base 3 is equivalent to 16 in base 10. The square root of 16 is 4. In base 3 this is 11. So the square root of 121 in base 3 is 11. What is the square root of 121 in the following bases: 4 5 6 7 8 9 10. What do you notice? Does this hold for *any* base?

A5

* *1* If $\mathscr{E}$ = {Letters of the English alphabet}
A = {Vowels in the English alphabet (excluding y)}
B = {Consonants (including y)}
W = {Words in the English language}
$X \in W$
L = {Letters in X}

Give an example of X such that

a) $L \cap B = \phi$ and $n(L) = 1$
b) $L \cap A = \phi$ and $n(L) = 2$
c) $n(L \cap B) = n(L) = 3$
d) $n(L \cap B) = 2$ and $n(L) = 3$
e) $n(L \cap A) = 2$ and $n(L) = 3$
f) $n(L \cap B) = 3$ and $n(L) = 4$
g) $n(L \cap B) = n(L \cap A)$ and $n(L) = 4$

2 The number 132 in base 4 has two factors, 11 and 12, also in base 4. Is this true for

a) base 5 *b*) base 6 *c*) base 7 *d*) base 8 *e*) base 9
f) base 10 *g*) any base?

3 Three quarters of the pupils in a class of about 30 have one dental filling or more, rather less than half have two fillings or more, and one quarter have three fillings or more. The number of pupils with two fillings only is three quarters of the number of pupils with three fillings or more. How many are there in the class? Represent the facts on a Venn diagram.

4 A motorist is travelling along a straight road at 45 kilometres per hour. He passes telegraph poles placed at regular intervals of 25 metres.

a) How far does he travel in *i*) 1 minute *ii*) 1 second?
b) How long does it take to travel the distance between two adjacent telegraph poles?
c) If he now speeds up so that he passes one telegraph pole a second, at what speed is he travelling?

* **5** The angles between adjacent diagonals drawn from a vertex A of a regular polygon are 30°. How many sides has the polygon? In another polygon, the angles between adjacent diagonals from a vertex B are 20°. How many sides has the polygon? How many sides would it have if these angles were 15°?

A6

* *1* If $A = \{\text{All prime numbers between 1 and 100 (excluding 1)}\}$ find $n(A)$

a) in base 10 *b)* in base 8 *c)* in base 5, giving n in the base stated.

2 Using D to represent 10, E to represent 11 and F to represent 12, carry out the following calculations:

a) Change 542_7 to base 12
b) Multiply 23_9 by 35_8 and give the answer in base 12. (*Hint* Change the two given numbers to base 10 first)
c) Divide 1232 by 8 in base 13. Do a denary check.

3 Using a Farey lattice put the following fractions into order of magnitude, smallest first:

$\frac{1}{4}$ $\frac{2}{7}$ $\frac{3}{8}$ $\frac{3}{13}$ $\frac{4}{9}$ $\frac{1}{6}$ $\frac{4}{15}$ $\frac{4}{17}$

4 In this question, round answers upward where necessary.

a) If a kilometre of wire costs £75.00, what is the cost of
i) five metres of the same wire *ii)* 1 metre *iii)* 80 centimetres?

b) If a tonne of potatoes costs £160, what is the cost in pence of
i) 50 kilograms of the same potatoes
ii) 1 kilogram of the same potatoes?

c) If you were buying the quantities stated in *b) i* and *ii*, would you expect to pay the prices you have calculated? If not, in each case say why not, and give what you think would be a reasonable price.

5 *ABCDE* is a regular pentagon. Draw in all the diagonals (there are five of them) and calculate every angle in the figure.

7 Nets and Polyhedra

7A Nets

1 Draw the net for a cube of side five centimetres. Be very accurate in your drawing.

* 2 The net you drew in question 1 contained six squares. In how many different ways can you arrange these six squares so that they still form the net of a cube?

'Different ways' means ways that cannot be changed into one another by reflection or rotation. Thus the four arrangements in fig. 1 can all be changed into one another by reflection or rotation, and are therefore not different ways. The two arrangements in fig. 2 cannot be changed into one another by reflection or rotation and so are different ways.

All the six arrangements are nets of cubes, but there are only two different nets altogether. You can easily convince yourself of this by drawing the six nets and cutting them out. By turning them over and twisting them round you will find five are the same and one is different.

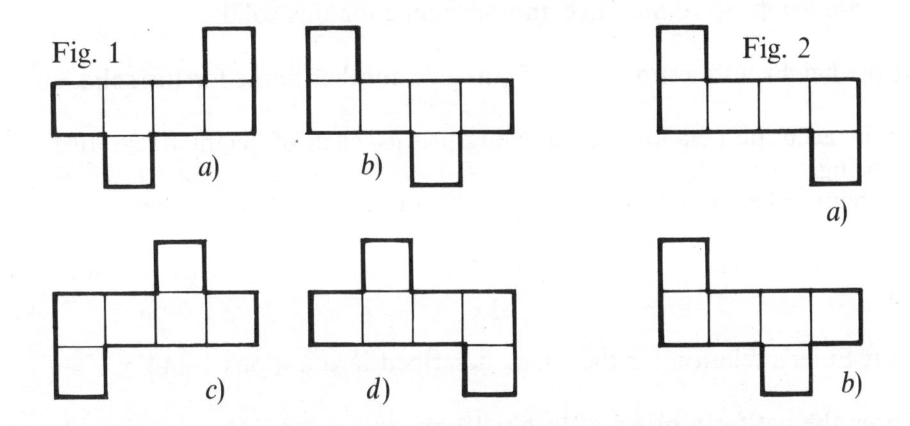

3 Construct the net of a cuboid of sides 6 cm by 4 cm by 4 cm.

* *4* How many different nets can you draw for question 3? (For the meaning of 'different', see question 2).

* *5* Repeat questions 3 and 4 for the cuboid 4 cm by 5 cm by 6 cm.

6 Take any one of the nets you have drawn so far and count F: the number of regions (including the outside region), E: the number of edges and V: the number of vertices. Check that Euler's relation, $F - E + V = 2$, is satisfied for the net you have chosen.

7 Repeat question 6 with one of the cubes or cuboids. This time F is the number of faces, and you do not add anything for the outside as this is not a face.

8 Use your net to calculate the total area of the faces of the cuboid in question 3.

9 Use your net to calculate the total area of the faces of the cuboid in question 5.

7B Prisms and Pyramids

1 Draw the net for a prism whose base is an equilateral triangle of side 5 cm and whose parallel edges are 10 cm.

∗ *2* In how many different ways can this be done? (See 7A, question 2 for the meaning of 'different ways'.)

3 Repeat question 1 for a prism whose cross section is a square of side 5 cm and whose height is 10 cm. Give another name for this solid.

4 If the height in question 3 was 5 cm, give another name for the solid.

5 Draw accurate nets for the following prisms, each of height 10 cm, their bases being

a) a regular hexagon of side 4 cm *b)* an octagon of side 4 cm.

6 Test Euler's relation, $F - E + V = 2$, for the nets you drew in questions 1 and 5.

7 Test Euler's relation for the solids described in questions 1 and 5.

8 Draw the net for a prism of height 10 cm, the base of which is the shape shown in the diagram.

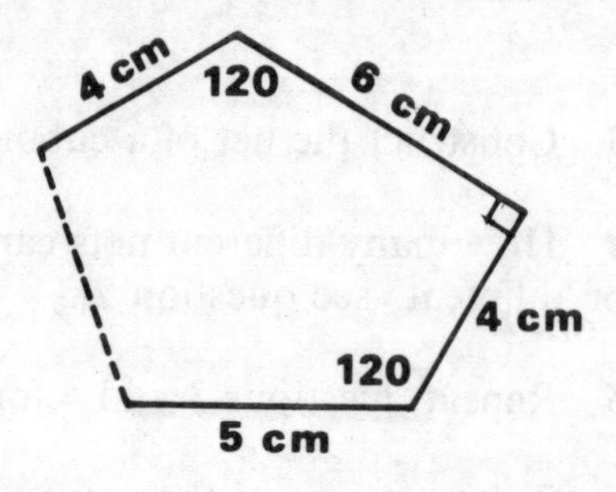

* 9 All the prisms in the previous questions are right prisms, i.e. the parallel sides are perpendicular to the base. In an oblique prism, the parallel sides are not perpendicular to the base.

The sides of a right prism are rectangles. What shape are the sides of an oblique prism? Are they all the same shape? Illustrate your answer with a freehand sketch.

** *10* Construct the following oblique prisms. (You may need some help.)

a) Base: an equilateral triangle of side 5 cm. One rectangular side inclined at 45° to the base. Parallel edges of length 8 cm.
b) Base: a regular hexagon of side 4 cm. Two opposite sides: parallelograms of base angle 60°, both perpendicular to base. Parallel edges of length 10 cm.

11 Draw the net for a pyramid whose base is a square of side 5 cm and whose edges are 8 cm. Construct this pyramid. Measure its height.

* *12* In how many different ways can you draw the net of this pyramid? For the meaning of 'different ways' see 7A, question 2.

13 Repeat question 11 for a pyramid whose base is an equilateral triangle of side 8 cm. Construct the pyramid and measure its height. This pyramid is a regular ...?

* *14* In how many different ways can you draw the net of the pyramid in question 13?

* *15* Repeat question 11 for
a) a pyramid whose base is a regular hexagon of side 4 cm and whose edges are 7 cm
b) a pyramid whose base is a regular pentagon of side 5 cm and whose edges are 6 cm.

16 Test Euler's relation for the nets you have drawn in questions 11 to 15, and also for the pyramids.

* *17* All the above are right pyramids with the vertex directly above the centre of the base. Now try and draw the net for a pyramid $VABC$ where the vertex V is directly above A and the base ABC is a right angled triangle with a right angle at A. Take AB and AC as 6 cm and VA as 7 cm.

* *18* If you have succeeded with question 17, try and draw the net of a pyramid $VABCD$ where the base $ABCD$ is a square of side 5 cm and the vertex V is 8 cm directly above A.

* *19* *a*) When is a pyramid, which is not a tetrahedron, regular?
b) What shape are the faces of any pyramid (excluding the base)?
c) What special feature is there about the faces of a right pyramid whose base is a regular polygon?

✱ 7C Other Polyhedra

1 Here are sketches of eleven different polyhedra. Below is a list of their names, the order being scrambled. State which name belongs to which figure.

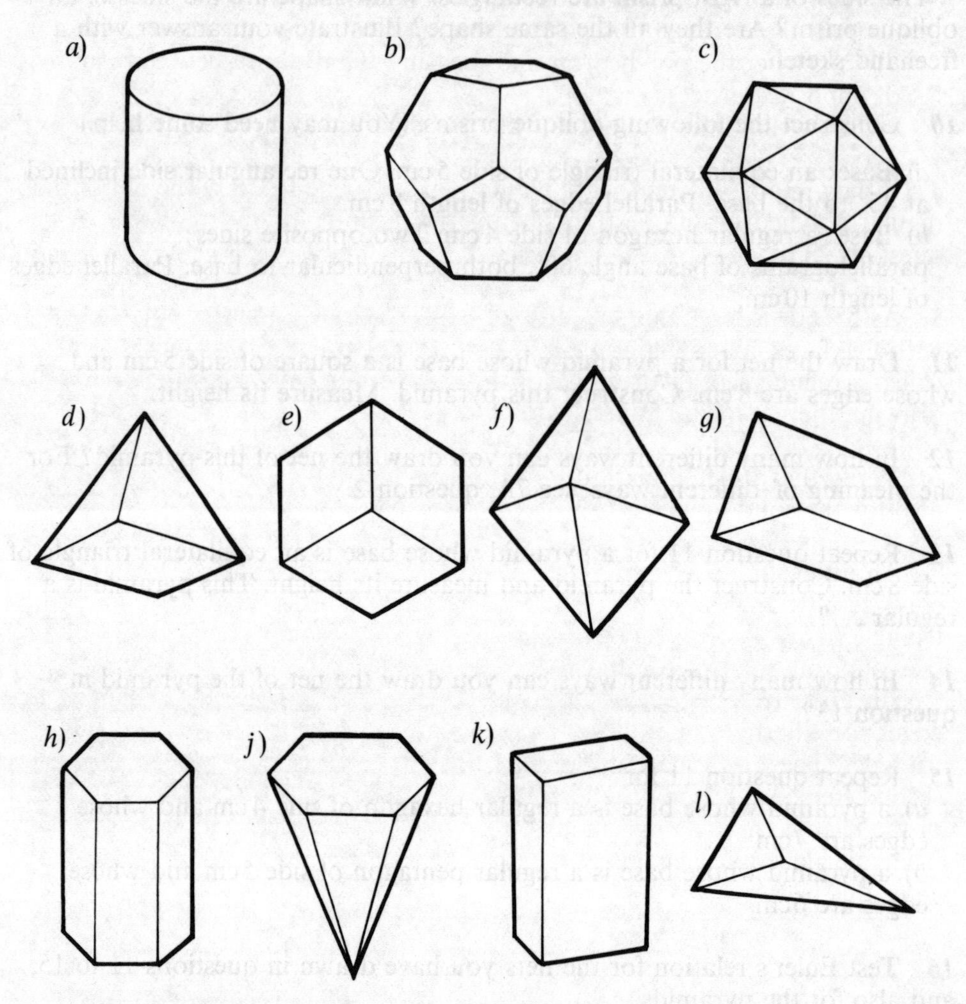

Scrambled list of names

i) triangular prism *ii*) tetrahedron *iii*) dodecahedron
iv) rectangular prism or cuboid *v*) cylinder *vi*) octahedron
vii) square pyramid *viii*) hexagonal prism *ix*) cube
x) icosahedron *xi*) hexagonal pyramid

2 The angles of an equilateral triangle are 60°. As $3 \times 60°$ is less than 360° it is possible to construct a regular polyhedron whose faces are equilateral triangles with three triangles meeting at every vertex.

What is the name of this solid? How many faces, edges and vertices does it have? Draw its net.

3 *a*) As $4 \times 60°$ is also less than $360°$ it is possible to construct a regular polyhedron whose faces are equilateral triangles with four faces meeting at each vertex. What is the name of this solid?
b) How many faces, edges and vertices does it have?
c) If all the edges of this polyhedron are 6 cm, draw the net.
d) Construct the polyhedron and measure the distance between any pair of opposite vertices.
e) Show that both the net and the polyhedron obey Euler's relation.

4 In how many different ways would it be possible to draw the net of the regular polyhedron in question 3? (For the meaning of 'different' see 7A question 2.)

5 As $5 \times 60°$ is less than $360°$ it is possible to draw a regular polyhedron with five faces, all equilateral triangles, meeting at each vertex. This solid is called a regular icosahedron. Find from any source you can a method of drawing the net for a regular icosahedron.

6 Verify that Euler's relation holds for the net of a regular icosahedron and also for the solid figure.

** **7** An interesting way of drawing the net in question 5 is as follows. Five equilateral triangles are to meet at each vertex. Start with one equilateral triangle, ABC, and on each side draw a regular hexagon. Omit the triangular face opposite to AB, BC, and CA.

This gives 16 faces out of the required twenty. There are six remaining free edges: on alternate ones draw three triangles. These will mate up with a second free edge leaving the third edge of the triangle free. So one more triangular face must be added to close this last space.

Note Free edges *a*, *b*, *c*, *d*, *e*, *f*.
Triangles drawn on *a*, *c*, *e*.
New free edges *g*, *h*, *i*, *k*, *l*, *m*.
bg, and *dk*, *fi* mate.
Closing triangle *lhm*.

Construct this net using triangles of side 5 cm. Make the solid and measure the length between two opposite vertices.

Note The angles of a square are $90°$. $3 \times 90°$ is less than $360°$, so a regular solid can be constructed with three squares meeting at a vertex. This is a cube. For exercises on a cube see 7A.

No regular solid with more than 5 triangular faces or three square faces meeting at a vertex can be constructed, as $6 \times 60°$ is $360°$ and 4×90 is $360°$.

** *8* The angles of a regular pentagon are 108°. $3\times108°$ is less than 360° so a regular solid can be constructed with three regular pentagons meeting at a vertex.

The net of this solid is constructed as follows. Draw a regular pentagon. On each side draw another regular pentagon. This gives half the net. The second half is drawn the same way.

To construct the solid from the net, join corresponding edges of the two halves of the net with sellotape or hold the two halves together with rubber bands.

How many faces, edges and vertices does the solid have? What is its name? Confirm that both the net and the solid obey Euler's relation.

** *9* Construct the net and the solid in question 8 using a pentagon of side 6 cm. Measure the distance between a pair of opposite vertices.

Note All the above regular solids are convex, i.e. any two adjacent edges meeting at a vertex enclose an angle of less than 180°. There are no more regular convex solids, but there are regular solids which are not convex. An interesting example is described in question 10.

** *10* Construct a regular dodecahedron as in question 8. On each pentagonal face of side 6 cm, construct a pentagonal pyramid. To get the right length of the sides of this pyramid proceed as follows. Draw a pentagon identical to any of the faces. Produce both ends of each side till they meet another side, also produced. This will give you a five pointed star. Folding up the triangles you have formed to meet at a point above the centre of the pentagon will give you the right shape of pyramid. The finished solid is a stellated dodecahedron.

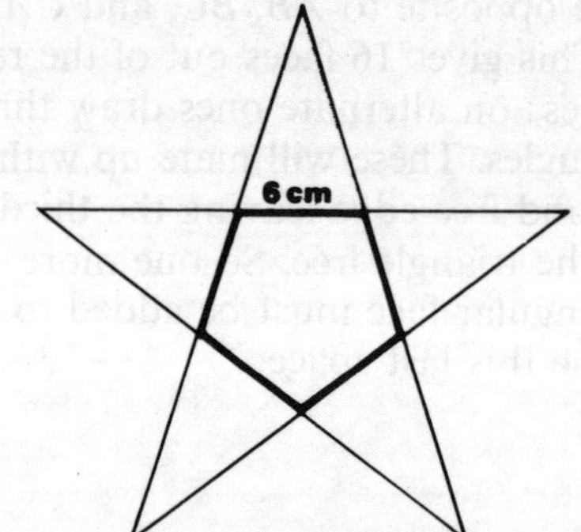

** *11* How many faces, edges and vertices does a stellated dodecahedron have? Does it obey Euler's relation?

** *12* With your completed model before you discuss the following questions.
a) The solid now has 60 faces. Why is it still called a dodecahedron? (The word 'dodecahedron' comes from the Greek and means a solid with 12 faces.)
b) Is it still regular?

* 7D Systematic Counting for Euler's Relation

Note The counting of edges, faces and vertices for solid figures is much simplified if a systematic approach is used.

1 Consider a cube standing on a table.

Faces There is a top and bottom face, and 4 upright faces.
Edges There are four in the top face and four in the bottom face. There are also four upright edges.
Vertices There are four in the top face, four in the bottom face and no others.

Hence calculate the total number of faces, edges and vertices and confirm Euler's relation for a cube.

2 Develop a similar systematic method of counting for

a) a square based pyramid
b) a tetrahedron
c) an octahedron

3 Consider a more complicated solid such as a dodecahedron. There are 12 faces, each with five edges. But two faces meet at each edge, so the total number of edges is $12 \times 5 \div 2$. There are five vertices in each face, but three faces meet at each point, so the total number of vertices is $12 \times 5 \div 3$. Check that the numbers of vertices, edges and faces calculated this way satisfy Euler's relation.

4 Apply the method of question 3 to *a*) an icosahedron *b*) a stellated dodecahedron *c*) a stellated icosahedron. (To make a model of *c*) construct a tetrahedron on each triangular face of an icosahedron.)

5 Count the number of edges and the number of vertices and faces in a cylinder. Is Euler's relation satisfied? Now draw a line on the curved face joining two points, one on the circumference of one end and one on the circumference of the other end. Test Euler's relation again. Explain your results.

7E Simple Volumes

1 Consider a cube of side 1 cm. Its volume is 1 cubic centimetre or $1\ \text{cm}^3$.

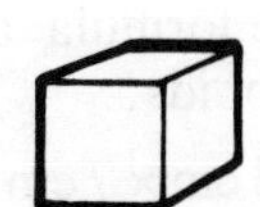

Now consider a cube of side 3 cm made up of layers of these little cubes.

a) How many little cubes are there in the bottom layer?
b) How many layers are there?
c) How many little cubes are there altogether?
d) What is the volume of the big cube?

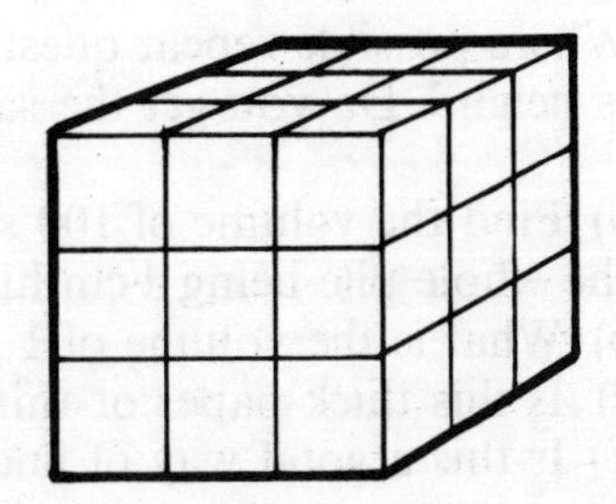

2 Thinking again about the cube of side 3 cm.

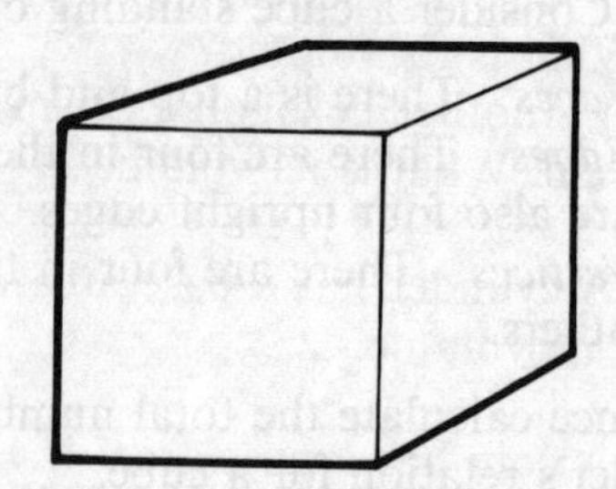

a) What is the area of the base?
b) What is the height?
c) What is 'area of base multiplied by height'?
d) Is this the same as the volume calculated in question 1*a*)?
e) Is it a general rule that the volume of a cube is 'area of base times height'?

3 Now think of a cube of side 4 cm, made up of little cubes of side 1 cm.

a) How many cubes are needed to fill the base layer?
b) How many layers are there?
c) What is the total number of little cubes?
What is the volume of the big cube?
d) What is the area of the base?
e) What is the height?
f) Calculate the volume as 'area of base times height'.
Does your answer agree with *c*)?

4 *a*) What is the volume of the cuboid in 7A, question 3?
Use the formula 'area of base times height'.
b) What is the volume of the cuboid in 7A, question 5?
c) Does it matter which side you take as base?

5 Consider a cuboid 6 cm by 7 cm by 8 cm.
a) Take the base as 6 cm by 7 cm. What is the area of the base?
What is the height? What is the volume?
b) Repeat *a*) taking the side 6 cm by 8 cm as base.
c) Repeat *a*) taking the side 7 cm by 8 cm as base.
d) Do all three calculations give the same volume?
e) Now use the formula 'length times breadth times height'.
Do you get the same volume?

6 Using the formula 'area of base times height', calculate the volumes of the following cuboids:

a) $4\text{ cm} \times 3\text{ cm} \times 7\text{ cm}$
b) $6\text{ cm} \times 6\text{ cm} \times 8\text{ cm}$
c) $2\text{ cm} \times 11\text{ cm} \times 1\text{ cm}$
d) $\frac{1}{2}\text{ cm} \times \frac{1}{2}\text{ cm} \times 100\text{ cm}$
e) base area 3 cm^2, height 18 cm

7 Where possible repeat question 6 using the formula 'length times breadth times height'. Do you get the same answers?

8 *a*) Find the volume of 100 sheets of paper, each sheet 30 cm by 18 cm, and the whole pile being 1 cm high.
b) What is the volume of 1 sheet?
c) Is this thick paper or thin paper?
d) Is this a good way of finding the volume of one sheet? Why?

9 *a*) A hexagonal prism has a base area of $16\,cm^2$ and a height of 4 cm. What is its volume?
b) Another hexagonal prism has a base area of $20\,cm^2$ and a height of 6 cm. What is its volume?

10 A cylinder has a base area of $22\,cm^2$ and a height of 5 cm. Find its volume.

11 A coin has a base area of $4\,cm^2$ and a thickness of $\frac{1}{4}$ cm. Find its volume. What have you neglected in this calculation?

12 *a*) Find the volume of a rectangular pile of 20 identical coins, the area of the face of each coin being $4{\cdot}8\,cm^2$ and the height of the pile being 7 cm.
b) Calculate the volume of one coin. Is your answer likely to be more or less exact than the answer in question 11? Why?

13 *a*) A cylindrical tin of base area $24\,cm^2$ holds water to a depth of 8 cm. Find the volume of the water.
b) Repeat for a tin of base area $30\,cm^2$ and a depth of 7·4 cm.
c) Repeat for another tin of base area $36\,cm^2$ and height 11 cm, the tin being half full.

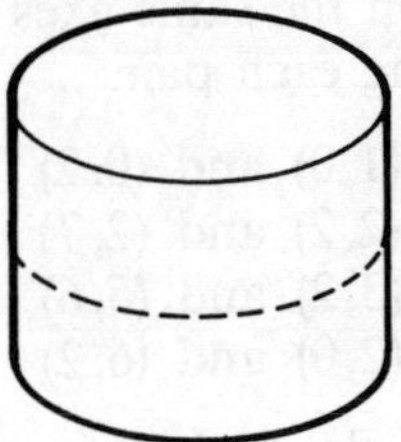

8 Co-ordinates, Graphs and Regions

8A Co-ordinates

1 On the same axes, plot these pairs of points and draw in the straight lines joining each pair:

a) (1, 0) and (3, 4) — (0, 3) and (4, 1)
b) (1, 4) and (4, 7) — (1, 7) and (5, 5)
c) (3, 2) and (5, 4) — (3, 5) and (6, 2)

Write down the co-ordinates of the points of intersection of each of the pairs of lines.

Two of these pairs of lines intersect each other at right angles. Which are they?

One of these two sets of 4 points if joined correctly will give a square. Why does the other set not give a square?

2 On the same axes plot these pairs of points and draw in the straight line joining each pair.

a) (1, 0) and (0, 2) *b)* (2, 1) and (0, 5)
c) (2, 2) and (2, 7) *d)* (3, 6) and (5, 8)
e) (3, 2) and (5, 6) *f)* (4, 3) and (6, 3)
g) (2, 0) and (6, 2)

Write down the co-ordinates of the midpoint of each of these lines.

3 Each of these sets of points form the corners of a square, but in each set the co-ordinates of one vertex are missing. On the same axes plot all the given points and fill in the missing co-ordinates:

a) $A(1,0)$ $B(2,1)$ $C(\quad)$ $D(0,1)$
b) $P(2,3)$ $Q(4,4)$ $R(5,2)$ $S(\quad)$
c) $W(\quad)$ $X(3,4)$ $Y(3,7)$ $Z(0,7)$

4 On the same axes plot these pairs of points and draw the straight line joining each pair.

a) (0, 0) and (4, 8) *b)* (2, 1) and (0, 3)
c) (1, 3) and (2, 3) *d)* (5, 2) and (4, 4)
e) (3, 3·5) and (0, 5) *f)* (1, 5) and (5, 7)

Write down the co-ordinates of the points of intersection between pairs of these lines.

5 Plot these seven points:

$A(1,1)$ $B(1,3)$ $C(3,3)$ $D(0,6)$ $E(1,6)$ $F(2,6)$ $G(6,6)$

If three or more of these points lie on a straight line, draw in the line. How many such lines are there?

Write down the sets of co-ordinates of the points on each line.

Write down anything which you notice about the sets of co-ordinates.

6 Plot these nine points:

$A(0,0)$ $B(4,1)$ $C(5,2)$ $D(3,2)$ $E(2,4)$
$F(3,6)$ $G(1,4)$ $H(1,2)$ $I(0,5)$

Draw straight lines which pass through three or more of these points. For each of these lines in turn write down anything you notice about the co-ordinates of the points which lie on it.

7 Plot these ten points:

$A(0,1)$ $B(0,6)$ $C(1,3)$ $D(1,2)$ $E(3,1)$
$F(2,4)$ $G(3,4)$ $H(3,5)$ $I(5,6)$ $J(5,1)$

Draw straight lines which pass through three or more of them.
What is the relationship between the co-ordinates of the points which lie on each line?

8 On the same axes draw the lines $y = 6$ $y = 3x$ and $y = x$.
Write down the co-ordinates of the points of intersection.

9 On the same axes draw the lines $x = 1$, $x = 2y$ and $x + y = 6$.
Write down the co-ordinates of the points of intersection.

10 If $A = \{(x, y);\ y = 2\}$ $B = \{(x, y);\ y = x + 1\}$
$C = \{(x, y);\ x = 5\}$ find $A \cap B$, $B \cap C$, and $C \cap A$.

8B Co-ordinates and Equations

1 Add three more to each of these sequences of ordered pairs of numbers:

a) (0, 0) (1, 2) (2, 4)
b) (0, 2) (1, 3) (2, 4)
c) (0, 7) (1, 6) (2, 5)
d) (1, 0) (2, 1) (3, 2)
e) (0, 0) (3, 1) (6, 2)

2 Fill in the gaps in these sets of co-ordinates:

a) (2, 0) (3, 1) (, 2) (6,) (, 6)
b) (1, 4) (2, 3) (3,) (, 1) (5,)
c) (0, 3) (2, 5) (3,) (, 9) (8,)
d) (0, 0) (2, 1) (3,) (, 3) (7,)
e) (0, 0) (1, 3) (2,) (, 9) (5,)

3 Write down the equations of the lines through the points represented by the ordered pairs in each sequence in question 1.

4 Write down the equations of the lines joining each set of points in question 2.

5 Write down the co-ordinates of four points which lie on each of these lines:

a) $y = 4x$ *b)* $y = x+2$ *c)* $y = 8-x$ *d)* $y = \frac{1}{5}x$ *e)* $y = x-4$

6 Write down the equations of the lines on which these points lie:

a) (0, 1) (1, 3) (2, 5) (4, 9)
b) (1, 2) (2, 5) (4, 11) (5, 14)
c) (0, 6) (1, 4) (2, 2) (3, 0)
d) (1, 0) (2, 2) (3, 4) (5, 8)
e) (0, 2) (1, 5) (3, 11) (4, 14)

7 Write down the equations of the lines on which these points lie:

a) (0, 0) (1, 5) (2, 10) (3, 15)
b) (1, 1) (2, 5) (3, 9) (5, 17)
c) (0, 10) (1, 7) (2, 4) (3, 1)
d) (0, 0) (1, 1) (2, 4) (3, 9)
e) (0, 1) (2, 7) (4, 13) (5, 16)

8 On the same axes, draw these lines, having first found, the co-ordinates of at least three points on each line.

a) $y = x$ *b)* $y = 2x$ *c)* $y = 3x$ *d)* $y = \frac{1}{2}x$

What do you notice about the slope of each line?

9 On the same axes, draw these lines, having first found the co-ordinates of at least three points on each.

a) $y = x$ *b)* $y = x+1$ *c)* $y = x+2$ *d)* $y = x-1$

What do you notice about these four lines?

10 Draw all these lines on the same axes:

a) $x+y = 6$ *b)* $y = 5-x$
c) $y+x = 4$ *d)* $x = 3-y$

Put all these equations into a form similar to that of *a)*. What do you notice about these lines?

8C Regions

Questions *1–10* in this exercise refer to the first quadrant only.

1 Draw each of these on a separate small graph and represent the required region by shading.

a) $x > 0$ *b)* $x < 5$ *c)* $y > 1$ *d)* $y > x$ *e)* $y < 7$

f) $y < x+2$ *g)* $y > 5-x$ *h)* $y > 0$ *i)* $y < 3x$ *j)* $y > 2x+1$

2 $A = \{(x, y): x > 1\}$ $B = \{(x, y): y < 5\}$
On a small graph show by shading the region $A \cap B$.

3 $P = \{(x, y): y > 1\}$
$Q = \{(x, y): x < 6\}$
$R = \{(x, y): y < x\}$
Show by shading the region $P \cap Q \cap R$.

4 $L = \{(x, y): y < 8\}$
$M = \{(x, y): x > 0\}$
$N = \{(x, y): y > 2x\}$
Show by shading the region $L \cap M \cap N$.

5 $A = \{(x, y): y > 1\}$
$B = \{(x, y): y < 7 - x\}$
$C = \{(x, y): y < x + 1\}$
Show by shading the region $A \cap B \cap C$.

6 Plot the points (1, 1) (5, 1) and (5, 5). Join them up and make a triangle. Write down the region inside the triangle in the form of the intersection of 3 sets as in question 5.

7 Plot the points (2, 0) (7, 0) and (2, 5). Join them up and make a triangle. Write down the region inside the triangle in the form of the intersection of 3 sets.

8 Draw the line $y = 2x + 3$. Plot the points (1, 3) (2, 8) (5, 8) (3, 6). Which side of the line do they lie? Are they in the region $y < 2x + 3$ or the region $y > 2x + 3$?

9 Draw the line $y = 3x - 2$. Plot the points (2, 6) (4, 8) (1, 4) (3, 8). Are they in the region $y < 3x - 2$ or the region $y > 3x - 2$?

10 Draw the line $y = 7 - 2x$. Plot the points (0, 5) (3, 2) (2, 7) (1, 3). Shade the region $y > 7 - 2x$. Which of these points come inside the shaded region?

11 Draw the lines $y = x + 3$, $y + x = 3$.
Shade the region $y < x + 3$, $y + x > 3$.

12 Draw the lines $x + y = 6$, $y = x$.
Shade the regions $x + y < 6$, $y < x$.

13 Draw the lines $y = x + 1$, $y = x + 4$.
Shade the region $y > x + 1$, $y < x + 4$.

14 Draw the lines $y-x = 4$, $y+x = 4$.
Shade the region $y-x > 4$, $y+x < 4$.

15 Draw the lines $2y = x+4$, $y = 2x+6$.
Shade the region $2y > x+4$, $y < 2x+6$.

16 Draw the lines $y = x$, $y+x = 0$.
These two lines divide the plane into four regions which are:

a) $y > x$, $y+x > 0$ *b)* $y > x$, $y+x < 0$

c) $y < x$, $y+x > 0$ *d)* $y < x$, $y+x < 0$

Label each of the four regions on the graph.

17 Draw the lines $y = 2x+4$, $y+2x = 4$.
Label the four regions, as in question 16.

18 Repeat question 17 for the lines $y = 3x$, $y = x$.

19 Choose one point in each region of question 16 and show that its co-ordinates do really satisfy the relations stated, e.g. (3, 5) is in the first region, and $5 > 3$, $5+3 > 0$.

20 Repeat question 19 for the regions in questions 17 and 18.

8D Conversion Graphs

1 Given that 5 miles = 8 kilometres, draw a graph to show the relationship between miles and kilometres.
Using the graph, convert these distances to kilometres:

a) 12 miles *b)* 22 miles *c)* 37 miles *d)* 52·5 miles *e)* 63 miles

Convert these distances to miles:

f) 5 km *g)* 12 km *h)* 25 km *i)* 45 km *j)* 68 km

2 Speed restriction signs in different places show these speeds in miles per hour. What are their equivalents in kilometres per hour?

a) 10 mph *b)* 30 mph *c)* 40 mph *d)* 70 mph

Use the graph of question 1.

3 Given that 72 kilometres per hour is the same speed as 20 metres per second, convert these speeds to metres per second by using a graph.

a) 20km/h *b*) 45km/h *c*) 100km/h *d*) 120km/h *e*) 185km/h

Convert these speeds to kilometres per hour.

f) 8m/s *g*) 15m/s *h*) 28m/s *i*) 42m/s *j*) 48m/s

4 On a certain date the rate of exchange for £10 was quoted as 24.00 dollars. Draw a suitable graph and find the equivalent number of dollars to each of these amounts:

a) £8 *b*) £17.50 *c*) £32 *d*) £45 *e*) £52.50 *f*) £63

5 The rate of exchange, on the same date, was quoted as £10 to 133.30 French francs. From a suitable graph find the equivalent number of francs to these amounts:

a) £6 *b*) £18 *c*) £27.50 *d*) £32 *e*) £43

Convert these amounts of French francs to £:

f) 100fr *g*) 250fr *h*) 320fr *i*) 460fr *j*) 525fr

6 An examination paper was marked out of a total of 130. Using a graph convert these marks to percentages:

a) 120 *b*) 105 *c*) 97 *d*) 82 *e*) 70 *f*) 58 *g*) 41

7 Another paper was marked out of a total of 85. Convert these marks to percentages:

a) 80 *b*) 72 *c*) 65 *d*) 57 *e*) 40 *f*) 33 *g*) 28

8 The circumference of a circle is proportional to the radius. When the radius is 8 cm the circumference is 50·3 cm. Draw a graph showing the relation between circumference and radius, and from it read off the circumference of these circles:

a) radius 4 cm *b*) radius 3 cm *c*) radius 5·9 cm *d*) radius $1\frac{3}{4}$ cm
e) radius 6·4 cm.

Find also the radius of the circles with these circumferences:

f) 28 cm *g*) 35 cm *h*) 16 cm *i*) 44 cm *j*) 22 cm

9 Using the graph of question 8, find the circumference of circles whose radius is

a) 14 cm *b*) 11 cm *c*) 12·4 cm

Find also the radius of circles whose circumference is

d) 96 cm *e*) 87 cm *f*) 66 cm

It may be necessary to use half the given figure, read the graph, and then double the answer.

10 *a*) A car owner finds that it costs him £200 a year plus 3p per km to run a car. Calculate the cost of running this car 5000km in a year. Calculate also the cost of running the same car 10 000km a year and 15 000km a year.
b) Draw a graph to show the cost of running a car. Plot distance horizontally, 1,000km to 1cm. Plot cost vertically, £50 to 1cm. The three points you have calculated should lie on a straight line. From your graph read off the cost of running a car

i) 4000km/year *ii*) 7500km/year *iii*) 12 000km/year
iv) 13 800km/year *v*) 9200km/year

* **11** Using the calculations you made in question 10, find the cost in pence per km, when the car is run 3000km, 4000km, 5000km, 10 000km and 15 000km per year. Find also the cost per km for 8000 and 12 000km per year. Draw a graph showing the cost per km. Plot km along the bottom to the same scale as question 10. Plot cost per km vertically, 2cm to one pence.

* **12** The graph in question 11 was a curve, not a straight line; smooth it well, and use it to find the cost per kilometre for:

a) 6000km/year *b*) 7000km/year *c*) 8500km/year *d*) 12 500km/year
e) 13 800km/year

13 The cost of electricity per quarter on the domestic tariff in 1973 was 3.75p per unit for the first 72 units, and then 0.85p per unit afterwards. Calculate the cost of

a) 24 units *b*) 48 units *c*) 72 units *d*) 200 units *e*) 500 units
f) 800 units *g*) 1200 units

Using a suitable scale plot your answers on a graph, with the number of units horizontal and the cost vertical. Join up points *a*) to *c*) with a straight line, and *c*) to *g*) with another straight line. (If you cannot do this, there is a mistake in your calculation or your plotting.)

Your graph should look like this:

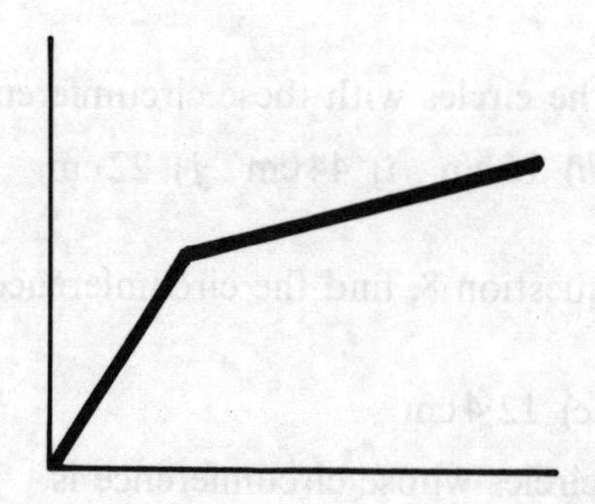

14 Use the graph in question 13 to find the cost per quarter of:

a) 50 units *b*) 100 units *c*) 400 units *d*) 670 units *e*) 1140 units

15 The cost of printing a pamphlet is made up of two parts. There is a fixed cost, and a cost that varies with the number of pamphlets. 500 pamphlets cost £28, 1000 pamphlets cost £38, and 2000 pamphlets cost £58.

Plot these figures on a graph, with the number of pamphlets on the horizontal axis and the cost in pounds on the vertical axis. From your graph find

a) the fixed cost
b) the varying cost per hundred pamphlets
c) the cost of 700 pamphlets
d) the cost of 1200 pamphlets
e) the cost of 1650 pamphlets.

* *16* A restaurant keeper decides to attract trade by offering a cut-price lunch. He finds that when he serves 30 lunches his overall profit is £2; on 50 lunches it is £6; and on 70 lunches £10.

Represent these figures on a graph. From your graph deduce

a) the least number of lunches he must serve to break even;
b) his loss if he sold only 10 lunches.

If on the average one customer in two bought a packet of cigarettes and the profit on these was four pence, draw a second graph above the first one to show the new situation.

c) How many lunches must he now serve to avoid a loss?

9 All about Numbers

9A Prime Numbers

1 Write down the numbers from 1 to 100 in ten rows of ten.

1 2 3 4 5 6 7 8 9 10
11 12 13 14 15 16 17 18 19 20
21 22 23 24 etc.

a) Cross off all the numbers divisible by 2.
b) Of the remaining numbers, cross out all those divisible by 3.
c) Of those still left, knock out all the numbers divisible by 5.
d) Finally knock out all the remaining numbers divisible by 7.
e) The numbers that are left are prime, i.e. they have no factors except themselves and 1. List the primes from 1 to 100.
f) Why did we not say 'cross out all the numbers divisible by 4' or 'by 6' or 'by 8' or 'by 9'?
g) Why did we stop at 7? Why did we not continue to 11, 13 etc?

2 Use the method of question 1 to find all the prime numbers between 100 and 200.

3 Use the method of question 1 to find all the prime numbers between 200 and 300.

4 The primes 11, 13 and 17, 19 in question 1 are known as prime pairs. Using your answers to questions 1, 2 and 3:

a) List all the prime pairs between 100 and 200.
b) Give the number of prime pairs between 100 and 200.
c) Give the number of prime pairs between 200 and 300.

5 The process described in question 1 is called the 'sieve of Eratosthenes'. To find whether a given single number is a prime by this method, you divide it in turn by all the primes smaller than its square root.

Use the sieve of Eratosthenes to decide whether or not 367 is a prime. Divide it in turn by all the prime numbers up to 19. (The next prime number, 23, is greater than the square root of 367.)

6 *Tests for divisibility*

1 If a number is divisible by 3, the sum of its digits is divisible by 3, e.g. 348 $\rightarrow 3+4+8 = 15$ (divisible by 3 so 348 is divisible by 3).
2 If a number is divisible by 5 it ends in 5 or 0.
3 If a number is divisible by 11, the number formed by adding and subtracting the digits alternately is also divisible by 11 (or is zero), e.g. 62535 $6-2+5-3+5 = 11$ (so 62535 is also divisible by 11).
4 The following test for divisibility by 7 or 13 is useful for numbers of 4, 5, or 6 digits. Write down the number formed by the last three digits and

also the number formed by the first 1, 2, or 3 digits. Subtract the smaller of these two from the larger. If the result is divisible by 7 or 13, so is the original number.
e.g. 461118 $461-118 = 343$, divisible by 7 but not by 13.
6175 $175-6 = 169$, divisible by 13 but not by 7.
Apply these tests to the following numbers:

a) Test 1 471; 1329; 268; 500
b) Test 3 4191; 52 318; 473; 654 324
c) Test 4 416 339; 8372; 29 815; 446 316

7 Use the method of question 5 to decide whether the following numbers are prime:

a) 371 *b*) 481 *c*) 487 *d*) 611 *e*) 911

If a number is not prime, give its factors.

8 Multiply 12 by 8. Add 1. Is the number prime?

9 Multiply 55 by 18. Add 1. Is the answer prime?

10 Multiply 23 by 20. Add 1. Is the answer prime?

∗ *11* If you multiply two whole numbers a and b and add one, and the resulting number is prime, what can you say about either a or b?

∗ *12* In question 11, is the answer always prime? Give examples.

13 Are the following numbers prime? If not, state their factors.

a) 629	*e*) 301	*i*) 871	*m*) 787	*q*) 881
b) 521	*f*) 977	*j*) 661	*n*) 793	*r*) 561
c) 523	*g*) 979	*k*) 461	*o*) 133	*s*) 469
d) 361	*h*) 727	*l*) 221	*p*) 433	*t*) 251

∗ *14* If there are L prime numbers between 1 and 100, M primes between 100 and 200, and N primes between 200 and 300, place L, M, N in numerical order, largest first.

∗ *15* Our arithmetic is calculated on base 10. What are the arguments in favour of changing to a prime base?

If you want to make mathematical history, try these:
a) Find a simple way of deciding whether or not a given number (a large number) is a prime, without slogging through the sieve of Eratosthenes.
b) Find a formula for the number of primes between two given numbers, e.g. 1200 and 1300.
c) Find a formula for the number of prime pairs between two given numbers, e.g. 1200 and 1300.
No one has yet found an answer to these questions.

9B Natural Numbers and Integers

1 List the natural numbers from 1 to 10.

2 If $A = \{\text{Natural numbers}\}$, what is $n(A)$?

3 List the integers from -10 to $+10$.

4 List the positive integers from 1 to 10.

5 List the negative integers from -1 to -10.

6 If $A = \{\text{Natural numbers}\}$ $\quad B = \{\text{Positive integers}\}$
$C = \{\text{Negative integers}\}$ $\quad D = \{\text{Integers}\}$

which of the following statements are true, and which are false:

a) $A = B$ *b)* $B \subset D$ *c)* $C \subset D$ *d)* If $\mathscr{E} = D$, $B' = C$

7 State a number which is in D but not in B or C. Write this result in set language.

9C Triangle Numbers and Square Numbers

1 Write down the first ten natural numbers in a column, one to a line. In a parallel column write down their cumulative sum.

Start of table

1	1
2	3
3	6

(6 is the sum of the first 3 natural numbers)

2 Write down the first ten natural numbers in a column, one on each line. In a parallel column write down the numbers obtained by adding one to each number in column 1. In column 3 write down the numbers obtained by multiplying the numbers in the first two columns. Compare column 3 with column 2 of question 1. What do you notice?

3 Calling the number on a given line of column 1 in question 2, N and the number on the same line in column 2, $N+1$, the number in column 3 is $N \times (N+1)$. Can you deduce a formula for the sum of the natural numbers from 1 to N?

4 The numbers 1, 3, 6 are called triangle numbers:

List the first ten triangle numbers.

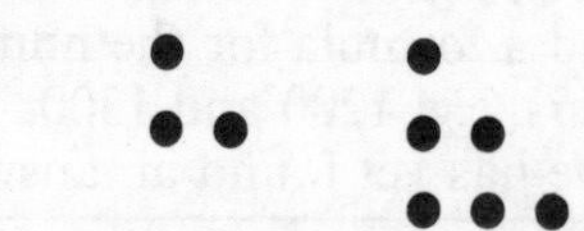

5 List the triangle numbers between 100 and 200.

6 Compare the triangle numbers in question 4 with column 2 of question 1. What do you notice? Can you explain it?

7 Write down the first 10 triangle numbers in a column, one to a row. In a second column write down the sum of each number and the one above it:

Start of table

1	1
3	4
6	9

What do you notice about the numbers in column 2?

8 Explain what you have observed in question 7 using the diagram in question 4.

* *9* Write down the first ten square numbers in a column, one on a line. In a second column write down their 'cumulative sums':

Start of table

1	1
4	5
9	14
16	30

(thus the figure in the second column against 9 in the first column is 14, which is the sum of 1, 4 and 9)

* *10* Write down the first ten natural numbers in a column (column 1).
In column 2 write down the numbers obtained by adding 1 to the numbers in column 1.
In column 3 write down the numbers obtained by *doubling* the numbers in column 1 and adding 1.
In column 4 write down the numbers obtained by multiplying the numbers in columns 1, 2 and 3. Compare column 4 with column 2 of question 9. What do you notice?

Start of table

1	2	3	6
2	3	5	30
3	4	7	84

* *11* Calling the number on a given row in the first column of question 10 N, the number on the same line of the second column is $N+1$, and in the third column is $2N+1$. Deduce a formula for the sum of the squares of the first N natural numbers. Check it for $N = 5$, 7 and 10.

9D Fibonacci Numbers

1 The sequence 1 1 2 3 5 8 13 ... is called a Fibonacci sequence. Each term is obtained by adding the two preceding terms. Give the next eight terms of the sequence.

2 Another Fibonacci sequence could be formed starting with 1 2 ....
Write down the first eight terms of this sequence. What do you observe? Can you explain this?

3 Write down the first ten terms of the sequence beginning 1 3 ....

4 Write down the first ten terms of the sequence beginning 1 4 ....

5 The numbers in the sequence in question 1 are given a special name. They are known as Fibonacci numbers. Write down the first twenty Fibonacci numbers.

6 If you write down enough Fibonacci numbers, the last digit starts repeating in the same order as the last digit of the first few numbers. At what term does this occur? To answer this question, it is only necessary to record the last digit of each number, not the whole number. This last digit is obtained by adding the two preceding last digits and discarding any 'tens' figures, e.g. starting with the fifth number 5

Index	*Last digit of number*
5	5
6	8
7	3
8	1
9	4

7 Write down the sum of the first three Fibonacci numbers. Compare it with the fifth number.
Write down the sum of the first four Fibonacci numbers. Compare it with the sixth number.
Repeat for the first 5, 6 . . . 10 Fibonacci numbers. What do you observe?

Note Further interesting properties of Fibonacci numbers are studied in Book 2.

0 Area

10A

In questions 1–5 use the side of one large square of your graph paper as 1 unit of length and the area of 1 large square as 1 unit of area.

1 Plot the vertices of each of these five figures, join them up and find the area of each by counting the number of squares enclosed.

a) (0, 7) (2, 7) (2, 9) (0, 9)
b) (0, 3) (2, 3) (2, 6) (0, 6)
c) (3, 3) (5, 3) (3, 6)
d) (0, 0) (2, 0) (1, 3)
e) (3, 0) (5, 0) (6, 3)

What do you notice about the area of *c*) in relation to the area of *b*)?
What do you notice about the areas of the three triangles?

2 Find the areas of each of the triangles formed by joining these vertices:

a) (0, 7) (3, 7) (3, 9)
b) (4, 7) (7, 7) (6, 9)
c) (0, 4) (3, 4) (1, 6)
d) (4, 4) (7, 4) (3, 6)
e) (0, 1) (3, 1) (5, 3)

What measurements are common to all five triangles?

3 Find, in the simplest way you can, the areas of the triangles formed by these points.

a) (2, 0) (6, 0) (5, 1)
b) (1, 2) (4, 2) (0, 5)
c) (5, 3) (7, 2) (7, 4)
d) (5, 4) (6, 5) (5, 6)
e) (1, 6) (4, 6) (3, 9)
f) (5, 7) (6, 7) (5, 9)

4 Draw the figures formed by these points. Find their areas by dividing them up into triangles and rectangles.

a) (1, 0) (0, 2) (1, 4) (2, 2)
b) (3, 1) (2, 3) (3, 4) (4, 3)
c) (4, 4) (6, 2) (6, 6) (4, 5)
d) (2, 5) (1, 7) (2, 8) (4, 7) (6, 5)

5 Find the areas of the parallelograms formed by joining these sets of points:

a) (2, 4) (4, 4) (2, 7) (0, 7)
b) (4, 6) (7, 6) (5, 8) (2, 8)
c) (0, 0) (2, 0) (3, 3) (1, 3)
d) (3, 0) (6, 0) (7, 2) (4, 2)

What can you say about the base and the height of these four parallelograms?

6 Calculate the areas of these shapes:

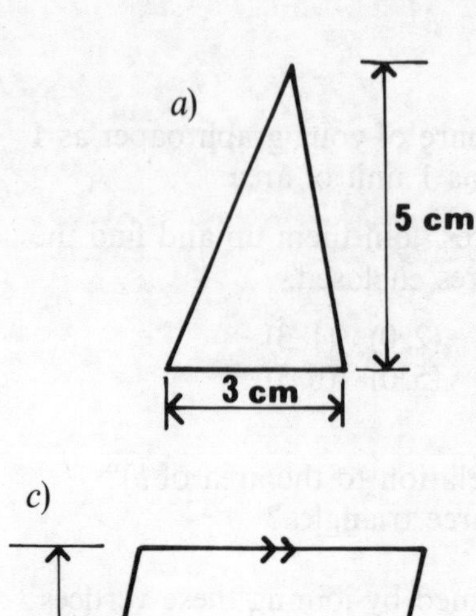

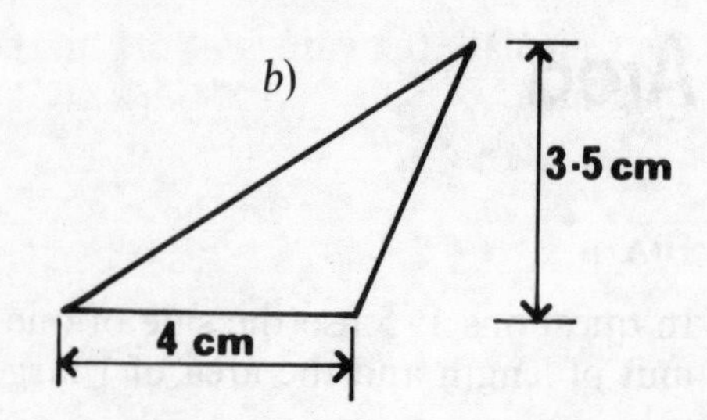

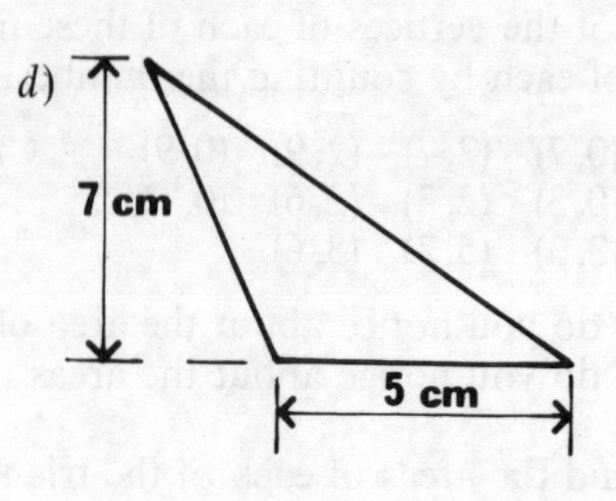

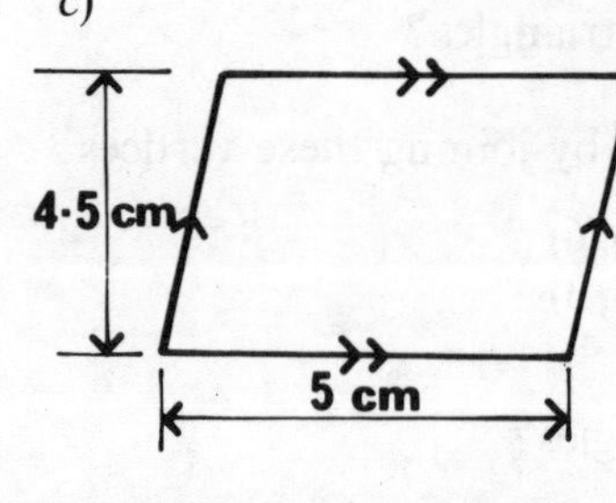

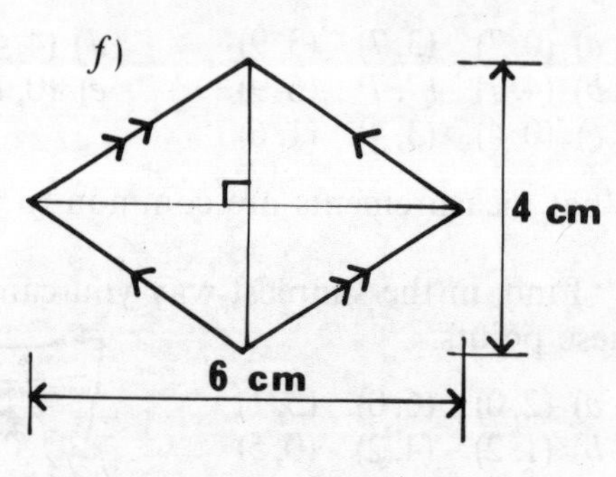

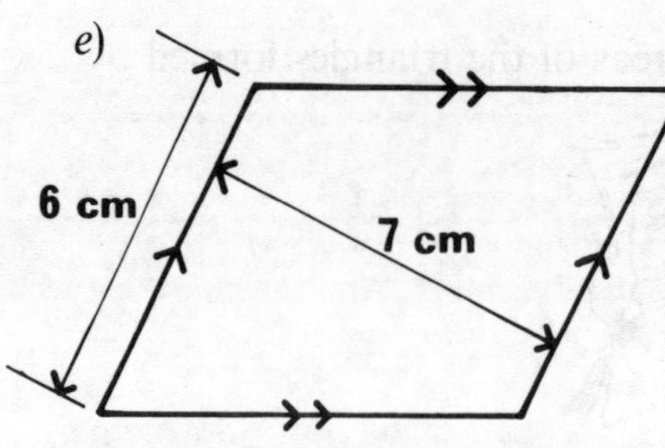

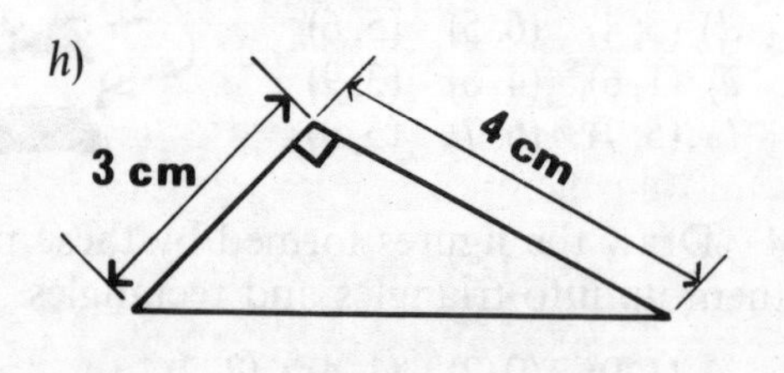

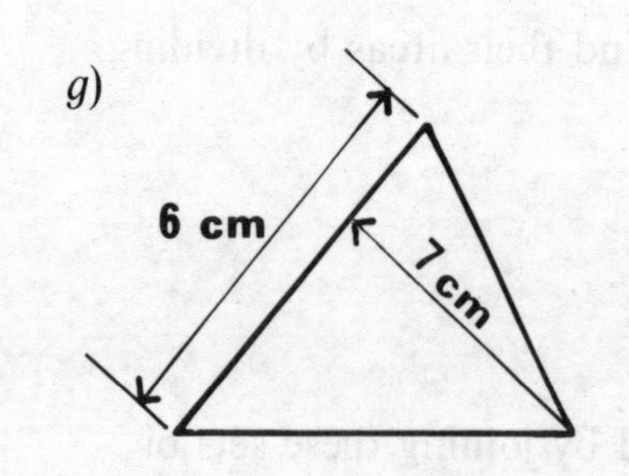

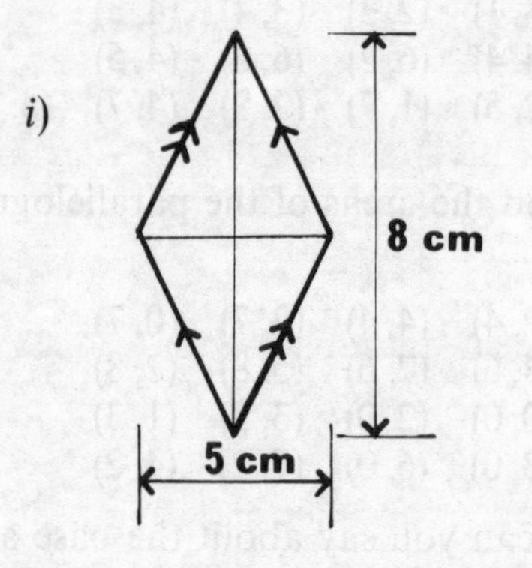

7 Calculate the values of the marked perpendicular heights in each of these diagrams:

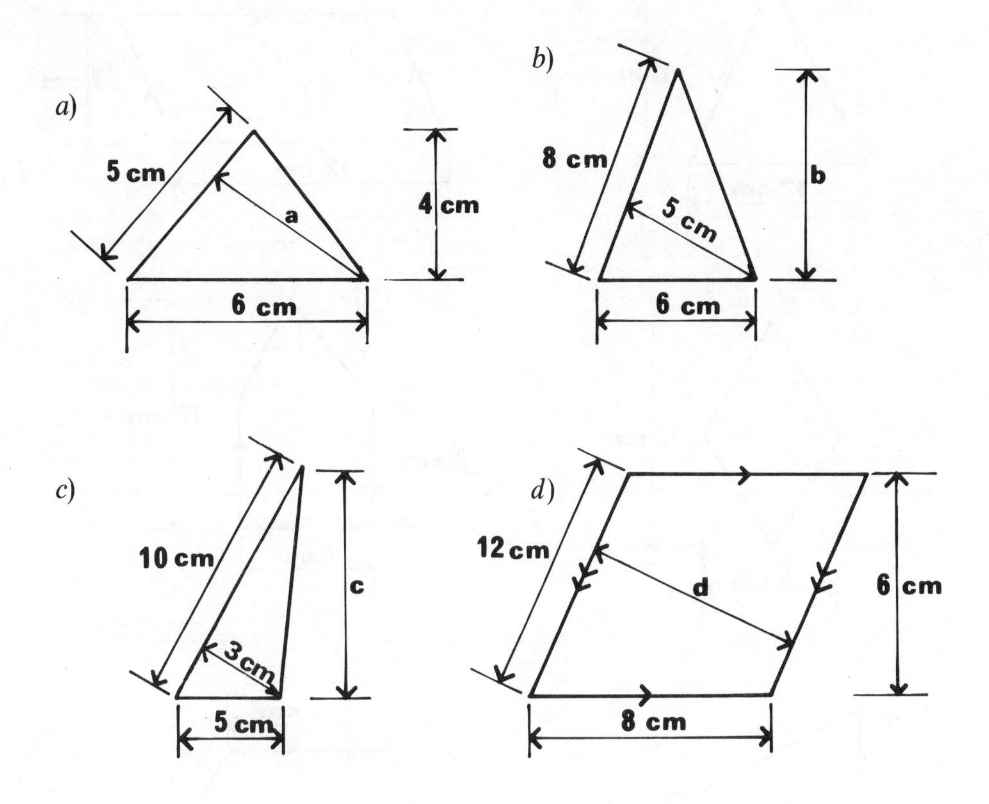

8 Using one large square to one unit on both axes, plot these four points and join them up to form a quadrilateral:

(2, 1) (6, 2) (5, 5) and (1, 3)

Find the area of the quadrilateral:

a) by 'boxing in', i.e. by surrounding it with a rectangle and subtracting the areas of the triangles at the corners, from the area of the rectangle.
b) by drawing a diagonal and dividing the quadrilateral into two triangles. Find the area of each and add the areas together.
Give your answer in both cases in square units.
Which method is the more accurate?

9 Repeat question 8 with a quadrilateral whose vertices are:

(3, 0) (6, 1) (4, 3) (0, 2)

10 Repeat question 8 with another quadrilateral whose vertices are:

(1, 0) (4, 2) (2, 6) (0, 5)

11 Find the areas of the following shapes:

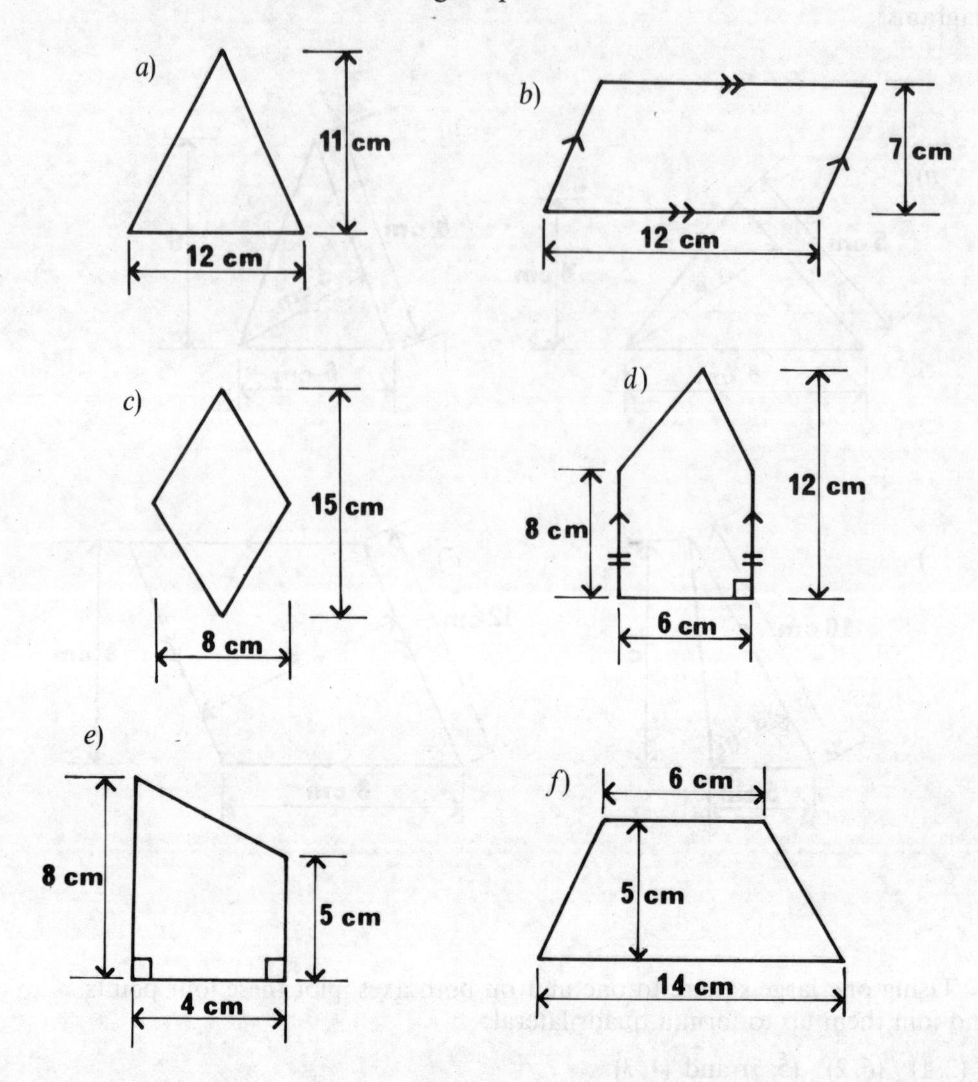

12 Calculate the lengths denoted by a and b if the area of the triangle is 30cm^2.

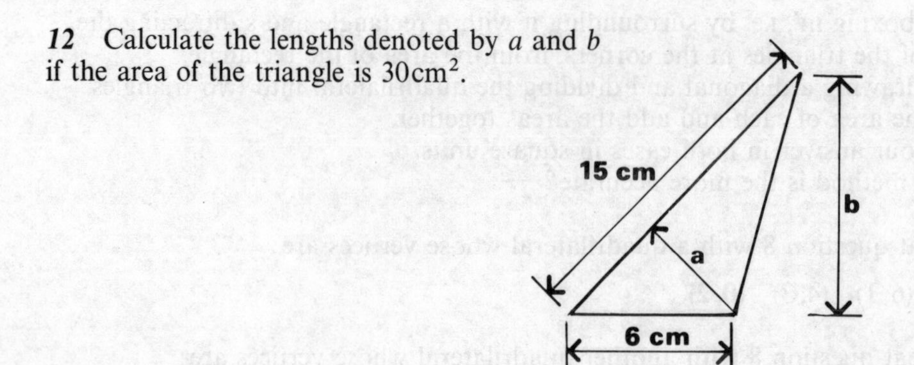

13 The figure shows the three altitudes of a triangle. They all meet at a point. If A is the area of the triangle, answer the following questions:

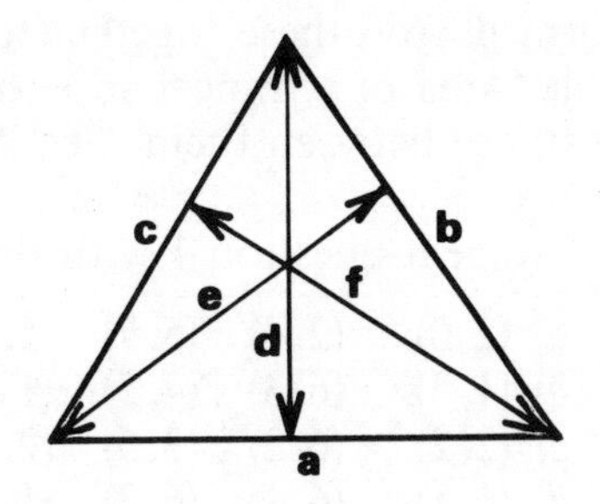

i) $A = 24\,\text{cm}^2$, $a = 6\,\text{cm}$, find d
ii) $A = 30\,\text{cm}^2$, $b = 8\,\text{cm}$, find e
iii) $A = 20\,\text{cm}^2$, $c = 10\,\text{cm}$, find f
iv) If $a = 6\,\text{cm}$, $d = 4\,\text{cm}$, $b = 3\,\text{cm}$, find e
v) If $e = d = 7\,\text{cm}$ and $A = 21\,\text{cm}^2$, find a and b. What kind of triangle is it? Is this always the case if $e = d$?

14 If A is the area of the parallelogram opposite, answer the following questions:

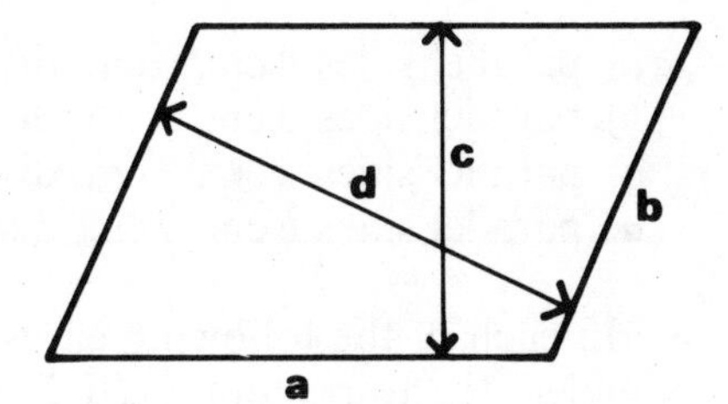

i) Find A if $a = 6\,\text{cm}$, $c = 4\,\text{cm}$
ii) Find d if $b = 5\,\text{cm}$, $A = 35\,\text{cm}^2$
iii) If c is twice d, what can you say about a and b?
iv) If a and b are equal, what about c and d? What is this particular figure called?
v) If $A = 18\,\text{cm}^2$ and a is twice c, what is c?

15 Construct the following triangles accurately to scale and find their areas. In each case do this in two ways. First select one side as 'base' and measure the corresponding altitude. Then select another side as base and measure the new altitude. Your two answers should agree quite closely:

a) sides 4 cm, 5 cm, included angle 60°
b) sides 3 cm, 4 cm, 6 cm
c) base 5 cm, height 6 cm, isosceles
d) base 6 cm, base angles 64° and 78°
e) sides 4·5 cm, 6·2 cm, included angle 120°

16 Construct the following parallelograms accurately to scale. Calculate their height and their area. In each case do this twice, selecting different sides as 'base':

a) sides 6 cm, 5 cm, base angle 50°
b) sides 7 cm, 4·8 cm, base angle 75°
c) sides 5 cm, 5 cm, base angle 45°
d) sides 5·4 cm, 6·2 cm, height (measured perpendicular to 5·4 cm side) 5 cm.

10B

1 Join the points (0, 0) (4, 0) (4, 3) (0, 5) to form a trapezium. Join the first and third of these points and calculate the areas of the two triangles formed. Add these together to get the area of the trapezium. Now apply the rule 'Area of a trapezium = half the sum of the parallel sides times the distance between them'. See if you get the same result.

2 Repeat question 1 with the following sets of points:

a) (2, 0) (2, 4) (5, 7) (5, 0)
b) (1, 3) (6, 3) (4, 5) (1, 5)
c) (2, 2) (6, 2) (3, 5) (2, 5)
d) (1, 1) (6, 1) (6, 3) (1, 7)
e) (4, 1) (10, 4) (6, 7) (2, 5)

Check the last one by boxing in (see 10A, 8*a*)).

3 Write down the areas of the following trapeziums:

a) parallel sides 4 cm, 6 cm, distance between them 5 cm
b) parallel sides 3 cm, 9 cm, distance between them 4 cm
c) parallel sides 2 cm, 5 cm, distance between them 7 cm
d) parallel sides 6 cm, 7 cm, distance between them 9 cm

4 In each of the following cases draw the quadrilateral *ABCD* and then complete the trapeziums *ABFE* etc. as shown. *i*) Calculate the areas of the trapeziums *ABFE* and *CBFH*. *ii*) Calculate the areas of the trapeziums *ADGE* and *CDGH*. *iii*) Find the sum of the two areas in *i*) and the sum of the two areas in *ii*) and take the second sum away from the first to get the area of the quadrilateral *ABCD*:

a) (2, 4) (4, 7) (7, 3) (5, 2)
b) (1, 2) (5, 5) (6, 1) (4, 0)
c) (0, 3) (2, 7) (5, 2) (2, 0)
d) (1, 2) (6, 3) (7, 1) (5, 0)

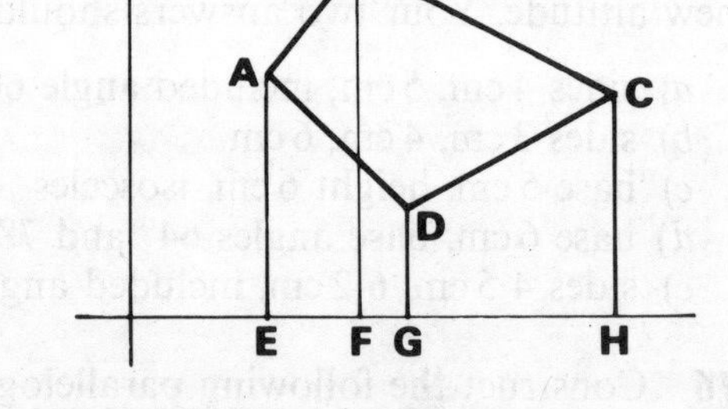

∗ Try also: *e*) (1, 3) (6, 1) (3, 4) (4, 6).
What must you do this time?
In each case, check by 'boxing in'.

5 Find the area of the figure formed by joining the points (2, 0) (5, 0) (7, 4) (4, 5) (0, 3): *a*) by boxing in *b*) by drawing lines parallel to the *y*-axis through (2, 0), (4, 5) and (7, 0) and finding the areas of two triangles and two trapeziums.

6 Draw the quadrilateral *ABCD* and calculate its area by any method you wish, given that:

a) $AB = 4$ cm, $BC = 5$ cm, $CD = 6$ cm, angle $B = 120°$, angle $C = 90°$
b) $AB = BC = CD = 4$ cm, angle $B = 60°$, angle $C = 140°$
c) $AB = 5$ cm, $BC = 4$ cm, diagonal $AC = 6$ cm, $AD = 5$ cm, $CD = 5$ cm

∗ *7* Draw a regular pentagon $ABCDE$ of side 6 cm. Each angle is 108°. Find the area of the pentagon *a*) by joining AD and BD and finding the areas of three triangles *b*) by joining D to F, the mid-point of AB, and finding the area of two trapeziums and two triangles (the trapeziums are formed by drawing lines through E and C parallel to DF).

∗ *8* Repeat question 7 for a pentagon of side 4·5 cm.

9 Draw a circle of radius 6 cm and inscribe in it a regular hexagon $ABCDEF$ of side 6 cm. Join opposite pairs of vertices to form six identical triangles all with their vertices at the centre. Find the area of one of these and hence the area of the hexagon. Check by joining AE and BD and finding the area of one rectangle and two triangles.

∗ *10* Repeat question 9 for a hexagon of side 5 cm.

∗ *11* Draw a regular octagon of side 4 cm (each angle is 135°). Find its area by any suitable method.

∗ *12* Repeat question 11 with a side of 5 cm.

10C Tessellations

1 Draw a square with sides 8 cm long. Mark off 2 cm lengths along all sides and join opposite points. This will give you a block of 16 squares, each with a side of 2 cm.

a) Along one edge add a row of similar 2 cm squares.
b) Along another edge add another row of 2 cm squares.
c) Along a third edge add another row of 2 cm squares.

This process can be continued indefinitely (except that you soon run off the paper). The squares 'tessellate' and fill the entire area of the plane.

2 Not all geometrical figures tessellate. Make a grid of dots at 2 cm intervals as shown. Taking each dot in turn as centre draw circles of 1 cm radius.
Shade the area between the circles. The circles do not fill the plane. The shaded area is left over. Circles do not tessellate.

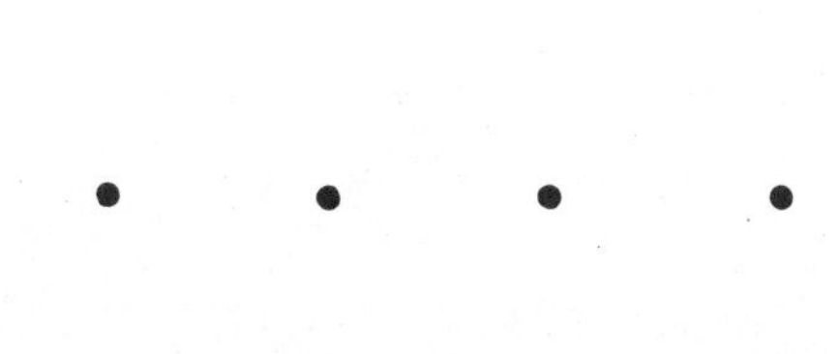

3 Discover for yourself by drawing, or by deduction, whether the following figures tessellate:

a) rectangles
b) equilateral triangles
c) isosceles triangles
d) any triangles
e) rhombuses
f) any parallelograms
g) trapeziums
h) regular hexagons
i) regular pentagons
j) regular octagons

4 Draw examples of the following figures *a*) that tessellate *b*) that do not tessellate. If it is impossible to do so, say so.

i) irregular pentagons *ii*) irregular hexagons *iii*) irregular quadrilaterals

5 Any figure that tessellates can be used as a unit of area. A square tessellates and is easily drawn, so a square of side 1 cm is the usual unit of area. We call its area 'one square centimetre'.

Draw an irregular quadrilateral to fill about half a page of your exercise book. Fill it with 1 cm squares. (In actual practice you can fill it with larger squares and rectangles, marking in each of these the number of 1 cm squares required to fill them.) You will be left finally with small triangular pieces round the edges. Count the total number of squares, estimating how many of the little triangles are equivalent to one square. This will give you the area of the figure in square cm.

6 In actual practice we do not count squares to find the area of a figure. One aim of mathematics is to find quick easy routines for avoiding tedious calculations. Several such routines have already been used in this chapter for finding areas, e.g. 'boxing in', dividing up into triangles, dividing up into trapezia, etc. Use any one of these methods to calculate the area of your quadrilateral and see if it agrees with the answer in question 5.

∗ 7 Draw a parallelogram of base 1 cm and height 1 cm. Its area is 1 cm^2. It tessellates. So it could be used as the unit of area in the same way as you used the unit squares in number 5. Would it be as convenient? Discuss.

∗ *8* Draw a pentagon as shown. Its area is 2 cm^2. It tessellates. Could you use it to measure areas as in question 5? Would it be as convenient as using the unit square? Discuss.

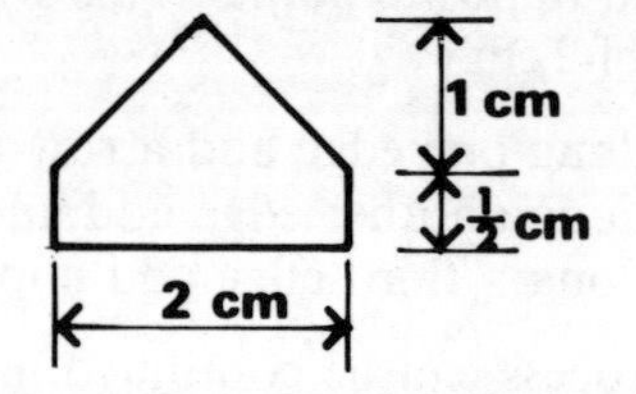

∗ *9* Find out what you can about the meaning of the word 'tessellation'.

11 Symmetry

11A Plane Figures

1 Each of these diagrams shows half of a symmetrical figure with the line of symmetry given as a dotted line. Copy these figures into your book and complete them:

a) 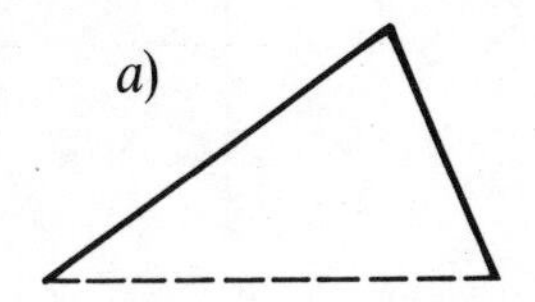

b) 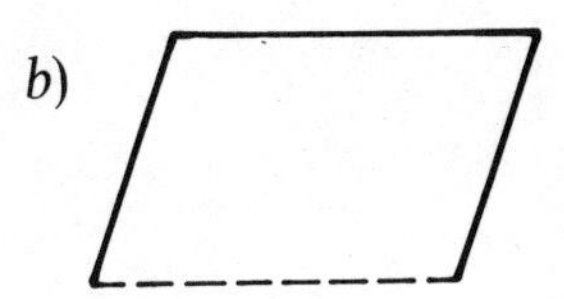

c) 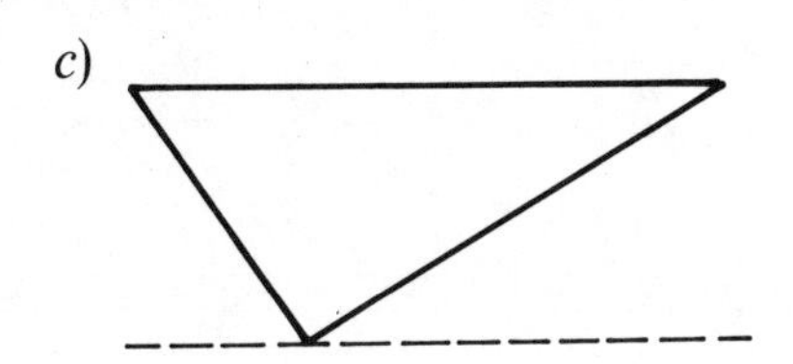

d) 

e) 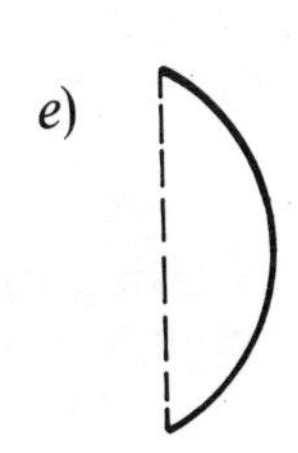

f) 

g) 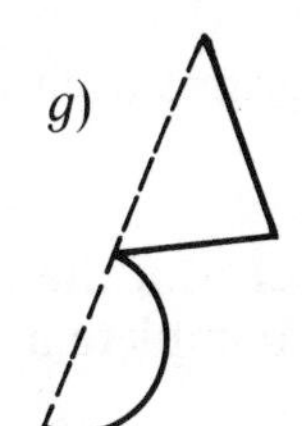

h) 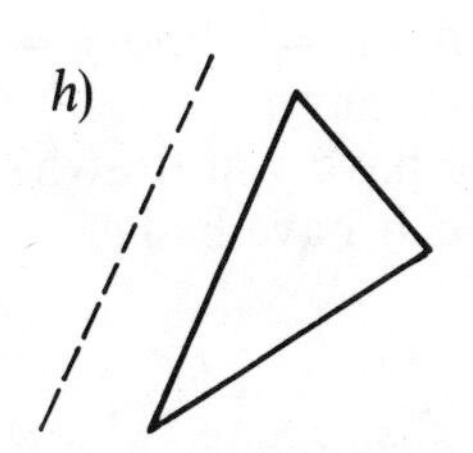

i) 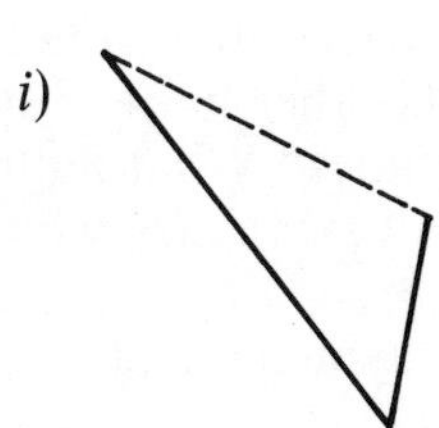

j) 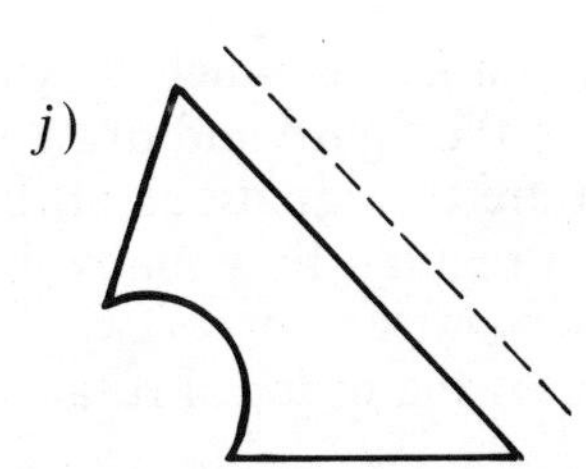

2 How many lines of symmetry can you draw into

a) a square
b) a rectangle
c) an equilateral triangle
d) an isosceles triangle
e) a parallelogram
f) a rhombus
g) a circle
h) a regular hexagon?

3 Draw five diagrams different from those above, each of which has one, and only one, line of symmetry.

4 Complete each of these diagrams so that each is symmetrical about the two lines of symmetry shown.

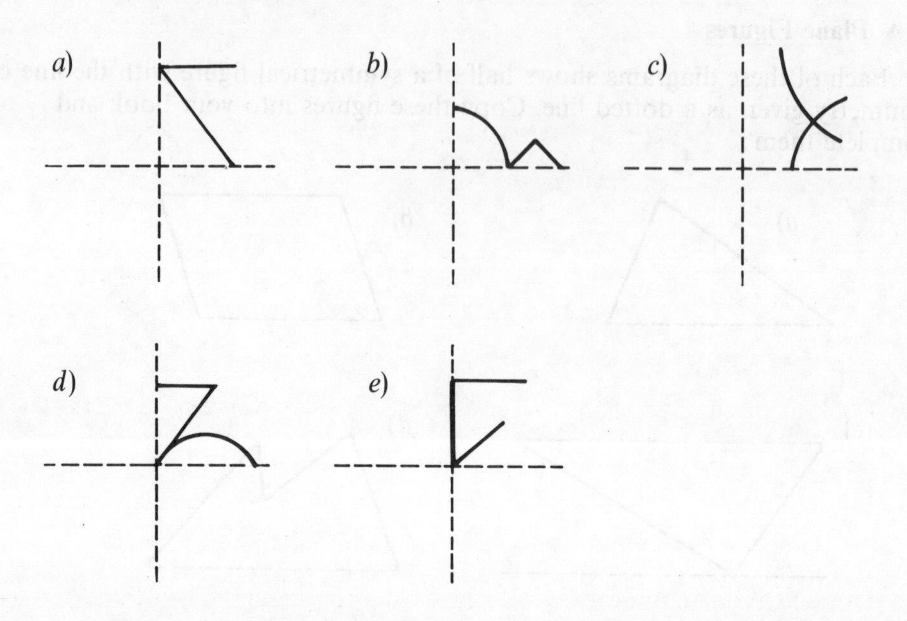

5 Draw five diagrams each of which has three, and only three, lines of symmetry.

6 On graph paper plot the points (0, 3) (0·5, 1) and (2, 2). Using these three points as vertices complete a figure which is symmetrical about $x = 0$, $y = 0$ and $y = x$.

Name another line which is a line of symmetry for this figure.

7 On graph paper plot the points (8, 2) and (6, 4). These two points are vertices of a figure which is symmetrical about the lines $x = 5$ and $y = 1$. Complete the figure and draw in its other lines of symmetry.

What are the equations of these lines? What figure have you drawn? If it had been regular, how many lines of symmetry would it have had?

8 What is the order of rotational symmetry of

a) a square *b*) a rectangle *c*) a rhombus
d) a parallelogram *e*) an equilateral triangle *f*) a regular hexagon
g) a regular octagon?

If you find any of these difficult to answer, cut out the shape and draw round it in your book. Fix the shape to your book with your compass point through the centre of rotation of both figures, and count how many times, in one complete revolution, the cut out shape fits the drawn shape.

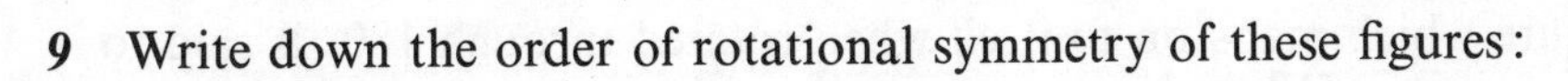

9 Write down the order of rotational symmetry of these figures:

a)

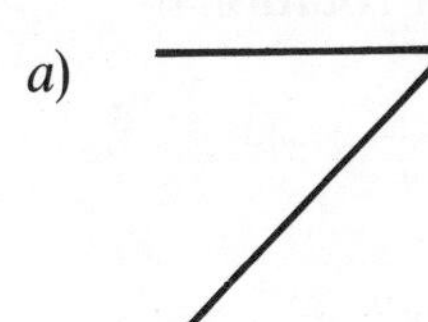

b)

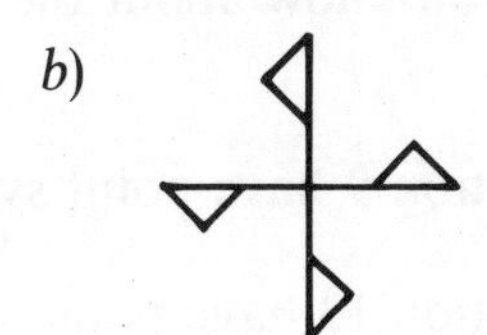

c)

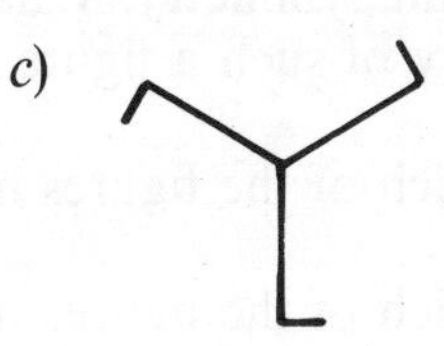

d)

e)

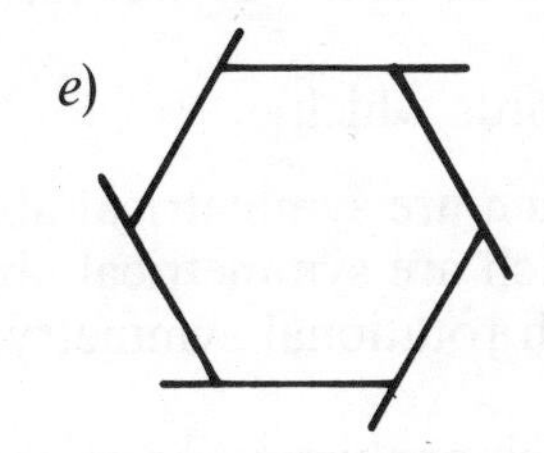

f)

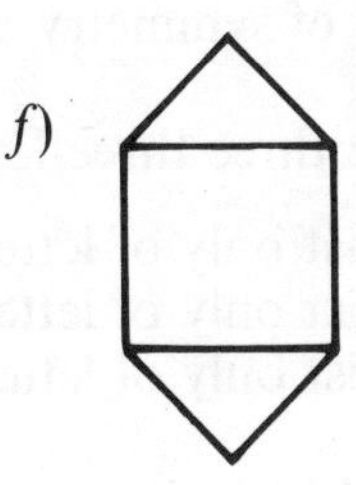

10 Look at the figures again in question 9 and write down how many lines of symmetry each has.

In those cases where the figure does not have line symmetry, copy the diagram and add to it to make the number of lines of symmetry the same as the order. Add to each diagram in the simplest possible way.

11 Draw for yourself figures which have no line symmetry, but have rotational symmetry of order

a) 2 *b)* 3 *c)* 4 *d)* 5 *e)* 6

12 Is it possible to draw figures which have line symmetry but not rotational symmetry?

13 Write down the number of lines of symmetry and order of rotational symmetry of each of these:

a)

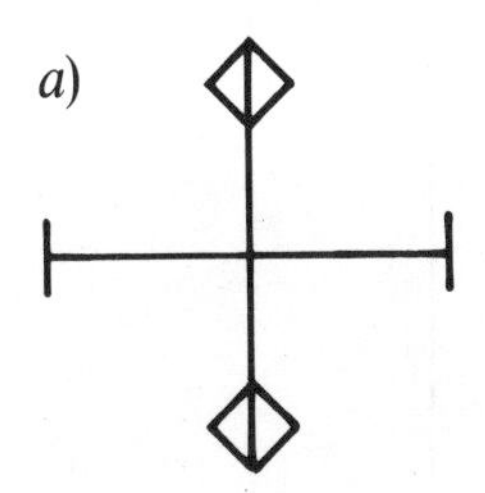

b)

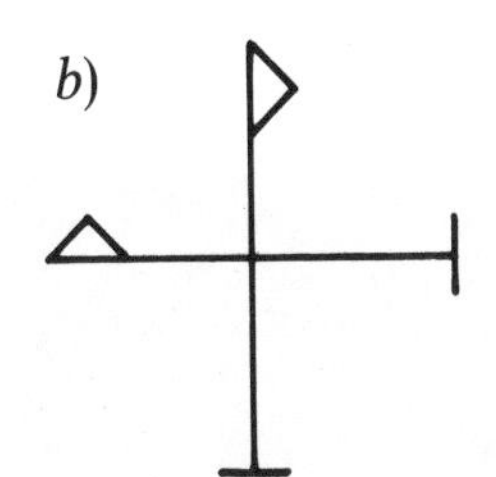

c)

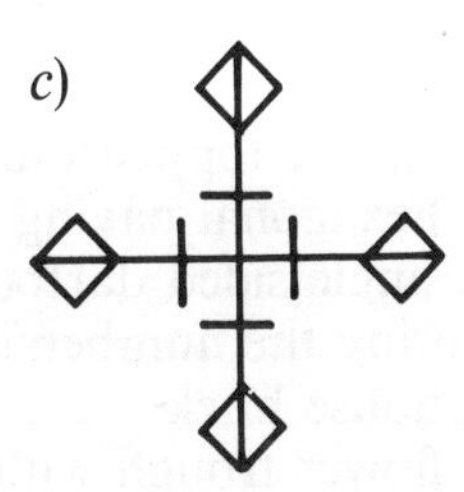

d)

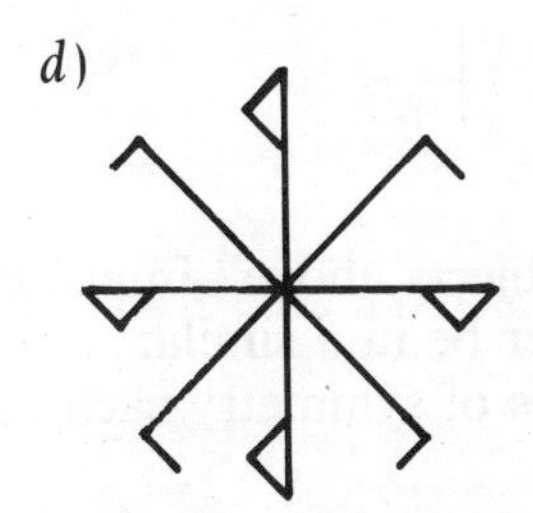

e)

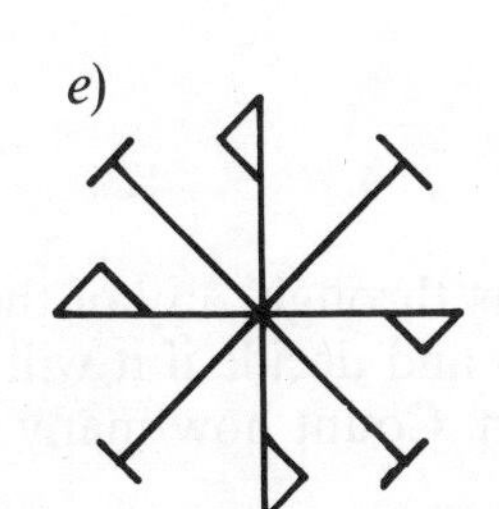

f)

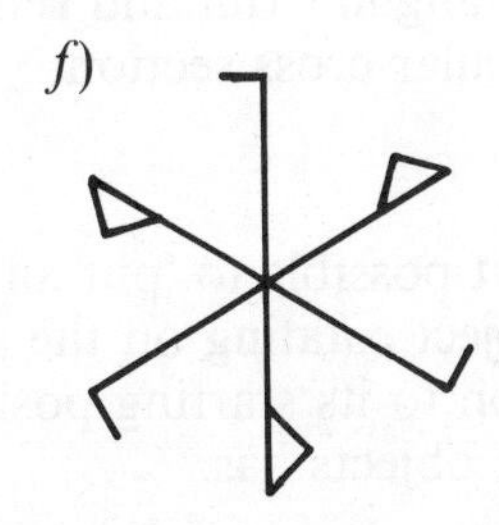

14 Any figure which maps on to itself when rotated through 180° is said to have point symmetry. What do you know about the order of rotational symmetry of such a figure?

15 Which of the figures in question 9 have point symmetry?

16 Which of the figures in question 13 have point symmetry?

17 Draw for yourself four figures with point symmetry, two of which also have lines of symmetry and two of which do not have line symmetry.

18 Write three three-letter words which

a) consist only of letters which are symmetrical about a vertical line
b) consist only of letters which are symmetrical about a horizontal line
c) consist only of letters with rotational symmetry.

19 List the playing cards which are symmetrical about

a) a horizontal line *b*) a vertical line

20 List the playing cards which do not have rotational symmetry.

11B Symmetry in Three Dimensions

Have as many models as possible in front of you while answering these questions.

1 Consider the symmetry of some 'everyday' objects and then complete this table:

	Planes of symmetry	*Order of symmetry*
a) A chair		
b) A square topped table		
c) A hexagonal paving slab		
d) A single sided dartboard (ignoring the numbers)		
e) A house brick		
f) A flower trough with a rectangular rim and semi-circular cross section		

2 Is it possible to 'put an axis through' any of the objects above? Imagine the object rotating on the axis and decide if it will ever be in a similar position to its starting position. Count how many axes of symmetry each of the six objects has.

3 Complete this table:

	Planes of symmetry	*Axes of symmetry*	*Order of symmetry*
a) A cuboid			
b) A prism whose cross section is an equilateral triangle			
c) A prism whose cross section is an isosceles triangle			
d) A cylinder			
e) A hexagonal prism			
f) A semicircular prism			
g) A square-based pyramid			
h) A hexagonal-based pyramid			

4 Complete the table for the five regular solids:

	Planes of symmetry	*Axes of symmetry*	*Order of symmetry*
Tetrahedron			
Cube			
Octahedron			
*Dodecahedron			
*Icosahedron			

Miscellaneous Examples B

B1

1

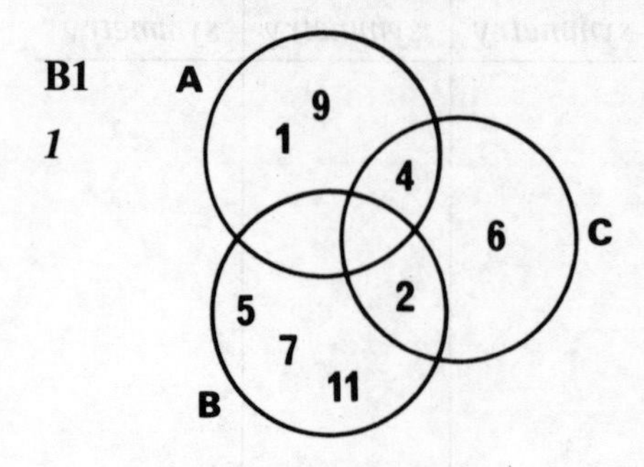

Write down possible descriptions for the sets A, B and C. Copy the diagram and mark the following numbers in the appropriate places: 16, 23, 25, 36, 56.

What can you say about $A \cap B \cap C$?

2 *a*) The total cost of knitting a sweater is £5.28, and a 50 g ball of wool costs 44p. How heavy is the finished garment if all the wool is used up?

b) The petrol tank of my car is half full. When I add 6 litres the petrol gauge shows that the tank is three quarters full. How much does the tank hold altogether?

3

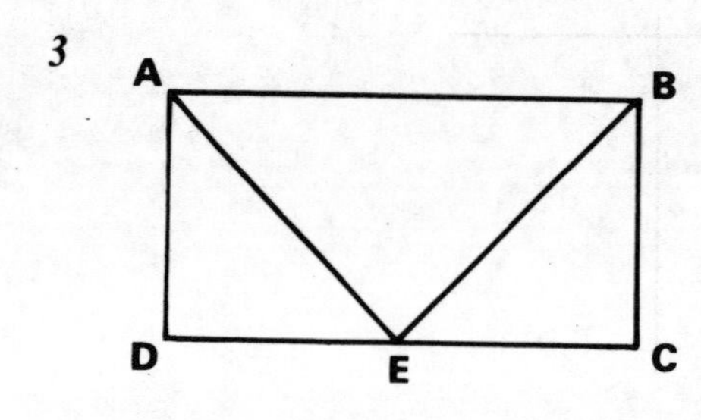

$ABCD$ is a rectangle in which $AB = 10$ cm and $BC = 5$ cm. E is the mid-point of DC. Find angle AEB and explain why triangle AEB is isosceles.

If F is a point on AE such that angle $ABF = 20°$, find angle FBE and angle BFE.

4 On graph paper, draw axes taking values from 0 to 10.
A cross has the following lines of symmetry:

$y = x$, $x+y = 10$, $y = 5$, $x = 5$. Draw these lines.

Part of the cross is formed by joining (5, 5) (6, 6) (7, 9) and (5, 9) to make a quadrilateral. Draw this shape and use the lines of symmetry to complete the cross.

Calculate the area of the complete cross.

5 How many faces has a tetrahedron? If a tetrahedron is painted so that no two adjoining faces are the same colour, what is the least number of colours needed?

If two tetrahedra of the same size are stuck together, face to face, how many colours are then needed if no two adjacent faces are the same colour?

B2

1 On graph paper, draw and label the lines $y = 1$, $y = 4$, $y = 4x$, and $x+y = 7$, using axes numbered from 0 to 8.

Show by shading the region which satisfies these inequalities:

$y > 1$, $y < 4$, $x+y < 7$ and $y < 4x$.

What shape is this region?

If $(4, p)$ and $(q, 3)$ lie inside this region, find possible values for p and q.

2 Express each of the following as the product of prime factors:

a) 120 *b*) 252 *c*) 209 *d*) 729 *e*) 1536

3 Find the height of an equilateral triangle of side 5 cm, by drawing the triangle as accurately as you can.

The net of a tetrahedron is cut from a piece of card 12 cm square. If rectangular tabs 5 cm by 1 cm are left on three edges, find the area of card wasted.

4 Multiply 26·5 by 4·3 giving your answer *a*) exactly *b*) correct to the nearest whole number *c*) correct to 1 decimal place.

Without further multiplication write down the exact answers to

d) 265×43 *e*) $2{\cdot}65 \times 4{\cdot}3$ *f*) $0{\cdot}265 \times 0{\cdot}43$

5 *a*) Add two more terms to the sequence 1, 4, 9, 16, ...

b) What is the tenth term?

c) What is the 20th term?

d) Add two more terms to the sequence 2, 5, 10, 17, ...

e) Write down the 10th and 20th terms of this sequence.

B3

1 Find the missing numbers so that each of the following addition sums are correct. They are all in the same base.

a)		*b*)		*c*)	
	103		41		3*2
	+ 42		+ 2*		+ 24*
	**0		1*0		1*30

What is this base?

2 Express each of the following as a product of prime factors:

a) 30 *b*) 210 *c*) 2310

Use your results to find the next two terms in the sequences

d) 2 6 30 210 2310 . . . *e*) 2 5 10 17 28 . . .

3 Ashfield, Barham and Canbury are three villages. Their positions are such that Canbury is 3 km from Ashfield on a bearing of 060°. The bearing of Barham from Ashfield is 028° and Barham is due north of Canbury.

a) By scale drawing, find the direct distance from Ashfield to Barham.
b) What is the bearing of Ashfield from Barham?

4 $A = \{\text{Points on the line } y = x-1\}$

$B = \{\text{Points in the region } y < 2x\}$

$C = \{\text{Points in the region } y > \frac{1}{2}x+1\}$

To which of these sets do the following points belong?

a) (1,1) *b*) (0,2) *c*) (2,1) *d*) (2,3) *e*) (4,3) *f*) (4,4) *g*) (5,4)

5 In a rectangle *ABCD*, *AB* = 8 cm and *BC* = 6 cm. *E* is the point in *AB* where *AE* = 2 cm. *F* is the mid-point of *DC*.

a) What is the area of *EBCF*?
b) What fraction of the area of *ABCD* is this?
c) How could you have deduced this fraction without calculating any areas?

B4

1 Draw a parallelogram of sides 6 cm and 4 cm and angle 55°. Draw in the two diagonals. From the point where they intersect, measure the perpendicular distance to each of the four sides. Calculate the areas of the four triangles. What do you notice? Could you have deduced this result without calculation?

2 *L* is the set of all points on the line $x+y = 6$, and *M* is the set of all points on the line $y = 3x$. Draw both these lines on the graph paper taking one large square as one unit. Mark $L \cap M$ and shade the region $x+y < 6$.

3 On a sheet of graph paper, using the same scale on both axes, mark the points (2,0) (4,0) (0,2) (0,6) (2,8) and (4,8). Is there a line of symmetry? If so, draw it in and give its equation. Add two more points and join up to form an octagon with two lines of symmetry. Draw in the second line of symmetry and give its equation. Mark the centre of symmetry and give its co-ordinates. What is the order of rotational symmetry?

4 *a*) Is 347_{10} a prime? If not, state its factors in base 10.
b) Is 347_9 a prime? If not, state its factors in base 9.
c) Is 347_8 a prime? If not, state its factors in base 8.

(You may find it easier in *b*) and *c*) to convert to base 10 and then convert back.)

5 A pyramid, vertex V, stands on a square base $ABCD$. V is directly above the centre of $ABCD$. The side of the square is 6 cm and the slanting edges are 9 cm. Draw an accurate net of the pyramid. From your net, find its total surface area.

B5

1 Look round your home and list any of the following objects that you can find: objects with order of symmetry 1, 2, 3, 4, 6, 8, 12. If you are ignoring printing, etc. say so. For instance, the order of symmetry of a 50p piece would be 14 if the printing and pattern were ignored.

2 If $\mathscr{E} = \{\text{Integers from 1 to 15}\}$

$A = \{1, 3, 7, 9, 13, 14\}$

$B = \{2, 3, 6, 8, 9, 12, 15\}$

write in full A', B', $A \cap B$, $A \cup B$, $A' \cap B'$, $A' \cup B'$.

Write down also $(A \cap B)'$, and $(A \cup B)'$. What do you notice?

3 Draw the polygon whose vertices are (3, 0) (5, 2) (7, 5) (2, 8) (0, 4). Find its area by 'boxing in'.

4 Simplify:

a) $2\frac{2}{3} - 3\frac{1}{4} + 1\frac{1}{2}$ *b)* $3\frac{1}{2} \times 1\frac{1}{7}$ *c)* $1 - \frac{1}{2} + \frac{1}{3} - \frac{1}{4} + \frac{1}{5}$

d) $1 - \frac{1}{2} - \frac{1}{4} - \frac{1}{8} - \frac{1}{16}$ *e)* $3\frac{3}{4} \div \frac{3}{4}$

5 On separate axes draw roughly the graphs of $y = x+2$, and $x+y = 6$, using a quarter of a page for each graph. Shade the regions $y > x+2$ and $x+y < 6$.

Repeat the graph using the same set of axes for both, and this time shade in the region in which $y > x+2$ and also $x+y < 6$.

B6

1 A regular octahedron has an edge of 4cm. Given that the height of an equilateral triangle is approximately 0·87 times the length of its base, find the area of one triangular face and hence the total surface area of the octahedron.

2 A firm in an outlying area employs semi-skilled labour at a weekly wage of £50. In addition the firm has to provide transport, and a 40-seater coach costs £200 a week. This cost does not vary whether the coach is full or only

half or a quarter full, but the proprietors refuse to allow standing passengers, and when the number to be transported exceeds 40, a second coach must be put on.

a) Work out the cost of wages plus transport for *i)* 10 workers *ii)* 20 *iii)* 30 *iv)* 40 *v)* 50 *vi)* 60 *vii)* 70 *viii)* 80.

Represent these figures on a graph, plotting number of workers horizontally and cost vertically. There will be a 'step' in the graph at 40 workers, i.e. the graph will jump up vertically.

b) From your graph read off the cost of wages plus transport for 25 workers 37 workers 41 workers 55 workers and 74 workers.

3 Multiply 23 by 12 in base 10 and get 276.
Is it true to say that 23 multiplied by 12 gives 276 in any of the following bases: 2 3 4 5 6 7 8 9? Where possible give brief reasons for your answer. Is it true for any other base?

4 Copy the following figures into your book and state for each the number of axes of symmetry and the order of rotational symmetry. Add not more than two lines to each figure to increase the symmetry and state for the altered figures the number of axes of symmetry and the order of rotational symmetry. Mark in the centre of symmetry.

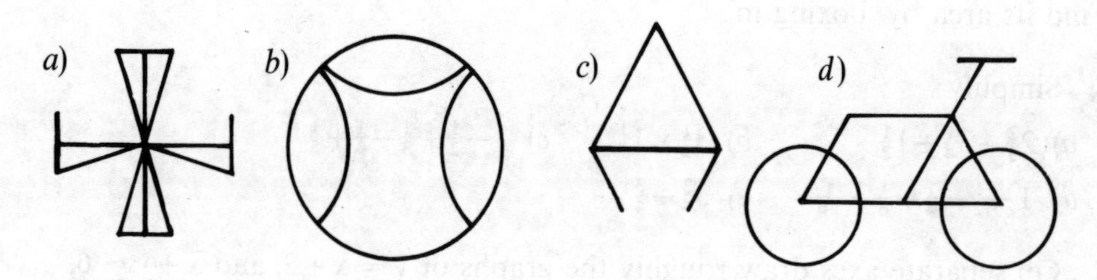

5 *a)* Change the following fractions to decimals:

$\frac{3}{16}$ $\frac{5}{8}$ $\frac{3}{4}$ $\frac{7}{16}$ $\frac{1}{3}$ $\frac{2}{3}$

b) Change the following decimals to fractions:

0·375 0·3125 0·85 0·6 0·825.

2 Algebra

12A

Find by inspection the value of a in each of these equations

1 $3a = 9$

2 $2a = 12$

3 $4a = 12$

4 $4a = 2$

5 $3a = 8$

6 $6a = 3$

7 $a+1 = 7$

8 $a+3 = 5$

9 $a+2 = 3$

10 $a-1 = 1$

11 $a-4 = 3$

12 $6-a = 2$

13 $7+a = 10$

14 $a-5 = 7$

15 $a+3 = 7$

16 $2+a = 9$

17 $a-4 = 6$

18 $a+4 = 6$

19 $a-10 = 2$

20 $10-a = 2$

21 $a-5 = 1$

22 $5-a = 1$

23 $a+3 = 3$

24 $a-4 = 4$

25 $a+8 = 12$

26 $\frac{1}{2}a = 5$

27 $\frac{1}{3}a = 2$

28 $4-a = a$

29 $6-a = a$

30 $\frac{2}{3}a = 8$

12B

Make up an equation to solve each of the following. Solve your equations. Do this by inspection in the simpler questions.

1 A number is doubled and 3 added, giving 11. Find the number.

2 When a number is subtracted from 15 the answer is 4. What is the number?

3 In 6 years' time Bob will be twice as old as he is now. How old is he now?

4 The distance round the perimeter of an equilateral triangle is 15 cm. What is the length of one side?

5 Mary spends three quarters of her pocket money on chocolate. If she spends 24p on chocolate, how much does her pocket money amount to?

6 Find the cost of one orange if I have 14p change from a 50p piece after having bought six oranges.

7 The sum of a number and twice that number is 21. What is the number?

8 The number which is five more than a certain number is 16. What is the number?

9 A number when trebled is the same as the number with 10 added. Find the number.

10 A rectangle is twice as long as it is wide. If the perimeter is 30 cm, what are its dimensions?

11 I think of a number, multiply it by five, add three and the answer is 28. What number did I start with?

12 I think of a number, multiply it by three then subtract five and find the answer to be seven more than the number itself. What is the number?

13 From a piece of string 20 cm long, two equal lengths are cut leaving a piece only half the length of one of the two lengths. What is the length of the remaining piece?

14 Mary is now twice as old as Jane. In three years' time the sum of their ages will be 21. How old is Jane now?

15 The sum of two consecutive numbers is 53. What are the numbers?

12C

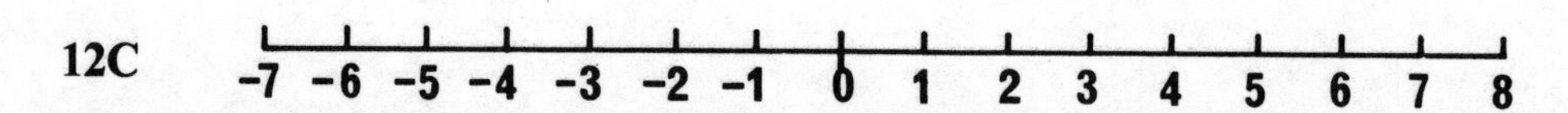

Use a number scale such as the one above to find the answers to the following. Remember that to the right is the positive direction and to the left the negative.

1	$+2-3$	*6*	$-2-3$	*11*	$-2-4$	*16*	$-4+7$
2	$-4+1$	7	$-1+7$	*12*	$+3+5$	*17*	$+5-8$
3	$-3-1$	*8*	$+5-5$	*13*	$+4-8$	*18*	$-5+8$
4	$+6-2$	*9*	$-3+0$	*14*	$-1-6$	*19*	$-6+6$
5	$+5+1$	*10*	$-5+6$	*15*	$+2-7$	*20*	$0+5$

12D

Write down the answers to the following, if possible without using a number scale.

1	$+2-1$	*6*	$-8+8$	*11*	$-10-3$	*16*	$+2-8$
2	$+3-2$	7	$-3+1$	*12*	$-1-4$	*17*	$-2-9$
3	$-2-5$	*8*	$-1-2$	*13*	$+3+4$	*18*	$+7-5$
4	$+3-6$	*9*	$+2-4$	*14*	$-2+5$	*19*	$-5+1$
5	$-9+6$	*10*	$-6+7$	*15*	$+3-7$	*20*	$-6-8$

12E

Write down the value of the following:

1	$-2-6$	*6*	$+10-7$	*11*	$-8-3$	*16*	$-3-7$
2	$+6+2$	7	$-10-7$	*12*	$-8+3$	*17*	$-3+4$
3	$-9+1$	*8*	$+7-10$	*13*	$+8-3$	*18*	$+8-7$
4	$+9-5$	*9*	$+7-7$	*14*	$-3+7$	*19*	$-8-7$
5	$-4+9$	*10*	$-8-8$	*15*	$+3-7$	*20*	$-9+5$

12F

You may find it helpful to use a number scale to find the answers to these subtraction sums.

1	$+5-(+2)$	*8*	$+7-(+6)$	*15*	$+6-(-2)$
2	$+6-(+4)$	*9*	$-2-(-1)$	*16*	$+3-(+5)$
3	$+3-(-1)$	*10*	$-4-(+2)$	*17*	$-5-(-1)$
4	$+4-(-2)$	*11*	$-6-(-6)$	*18*	$-3-(+2)$
5	$+4-(+2)$	*12*	$-1-(+3)$	*19*	$-7-(-8)$
6	$+8-(+3)$	*13*	$-2-(+5)$	*20*	$-4-(+5)$
7	$+7-(-2)$	*14*	$-3-(-6)$		

12G

Write down the answers to these subtraction sums:

1	$+3-(-4)$	*8*	$+5-(+8)$	*15*	$-4-(+7)$
2	$+3-(+4)$	*9*	$-9-(+10)$	*16*	$+8-(+6)$
3	$-3-(-4)$	*10*	$+9-(+10)$	*17*	$+8-(+10)$
4	$-3-(+4)$	*11*	$-4-(-8)$	*18*	$-6-(+1)$
5	$-7-(+1)$	*12*	$-2-(-6)$	*19*	$+1-(-4)$
6	$-7-(-1)$	*13*	$-2-(+7)$	*20*	$+1-(+5)$
7	$+5-(-8)$	*14*	$-5-(-2)$		

12H

1	Subtract $+5$ from $+7$	*4*	Subtract $+5$ from -7
2	Subtract -5 from $+7$	*5*	Subtract $+3$ from $+4$
3	Subtract -5 from -7	*6*	Subtract $+4$ from -6

7 Subtract -9 from -9

8 Subtract $+7$ from $+2$

9 Subtract -1 from $+4$

10 Subtract -1 from $+1$

11 Subtract $+3$ from -1

12 Subtract -9 from $+6$

13 Subtract -1 from $+2$

14 Subtract $+8$ from $+1$

15 Subtract $+3$ from -4

16 Subtract -3 from -4

17 Subtract -2 from -6

18 Subtract $+1$ from $+3$

19 Subtract $+2$ from -5

20 Subtract -8 from -3

12I

Solve these equations. The simpler ones you may do by inspection.

1 $5 = a-1$

2 $12-a = 7$

3 $2a+1 = 5$

4 $2a-3 = 5$

5 $13-2a = 1$

6 $9-a = a$

7 $15 = 4a+3$

8 $\frac{3}{4}a = 6$

9 $\frac{2}{5}a = 4$

10 $a+3 = 2a$

11 $5+a = 2a$

12 $3a-2 = a$

13 $12-a = 2a$

14 $a+1 = 2a$

15 $2a = 1-a$

16 $3a = 3-a$

17 $6+2a = 11$

18 $3a-2 = 14$

19 $5a = 3a+4$

20 $3a+2 = 4a$

21 $3a-1 = 2a$

22 $2a+1 = a+4$

23 $2a+3 = a+5$

24 $a+8 = 2a+5$

25 $2a-1 = a+1$

26 $2a-4 = a+3$

27 $2a-3 = a-1$

28 $a+7 = 3a-1$

29 $4a+5 = 2a+7$

30 $4a-3 = a+2$

12J

Solve the following equations:

1 $a-2 = 9$

2 $2-a = a$

3 $7+2a = 3$

4 $4a = a+6$

5 $5a-6 = 2a$

6 $4a+3 = 11$

7 $2a+7 = 4a$

8 $5+3a = 8$

9 $8-3a = 5$

10 $8a-3 = 5a$

11 $2a+9 = 5a$

12 $3a-6 = 0$

13 $4a-2 = 0$

14 $\frac{1}{5}a = 3$

15 $\frac{3}{7}a = 6$

16 $a+2 = 2a-1$

17 $3a+2 = a+8$

18 $4a+1 = 2a+3$

19 $2a+3 = 3a-5$

20 $5a-1 = a+1$

21 $4-a = 1+a$

22 $3+2a = 8-3a$

23 $5a-1 = 1$

24 $5a-1 = a$

25 $5-a = a$

26 $3-a = 3+a$

27 $2a+9 = 13-a$

28 $3a-7 = 7-a$

29 $8-2a = 10-3a$

30 $3-a = 8-3a$

12K

Solve the following equations:

1 $a+2=5$
2 $a-3=6$
3 $a+7=3$
4 $3+a=1$
5 $a=4-a$
6 $2a=a-6$
7 $5-a=2a$
8 $7-2a=a$
9 $3a+1=a$
10 $2a+5=3$
11 $\frac{2}{3}a=6$
12 $\frac{1}{4}a=5$
13 $6=3a+2$
14 $10+a=3a$
15 $5+a=1$
16 $a+4=4$
17 $3a+2=5$
18 $3a+7=1$
19 $1-4a=0$
20 $1+a=0$
21 $2a-3=-1$
22 $a+6=-2$
23 $2a-3=a+5$
24 $2a+4=1+a$
25 $3a+7=15-a$
26 $2a+3=6-a$
27 $4a+12=2-a$
28 $7-a=a-3$
29 $2a+3=3a+5$
30 $4a=2a-1$

12L

Simplify the following wherever possible:

1 $a+a+a$
2 $a+a+b$
3 $a\times a$
4 $a\times a\times b$
5 a^2+a
6 $a^2\times a$
7 $2a+a$
8 $2a^2+a$
9 $2a\times a$
10 $a-a$
11 a^2-a
12 $2a-a$
13 $a-b$
14 a^2-b^2
15 $a+2b+3a$
16 $2a\times 3a$
17 $a\times b\times 2$
18 $a+b+2$
19 $5a+3a$
20 $5a-3a$

12M

Simplify the following wherever possible:

1 $3x\times 4y$
2 x^2+2x+4
3 $(3x)^2$
4 $3y\times y$
5 $ab+ba$
6 $abc+bca+cab$
7 $ab+bc+ca$
8 $ab\times bc$
9 $3ab\times 5bc$
10 $(2x)^3$
11 $a+2b+3a$
12 $a-3b-2b$
13 $x+3y-x-2y$
14 $3x-2y+5y$
15 $4-2a+5b-3$
16 $2x^2-5x^2+7x^2-x^2$
17 $4a^2-3a+2a^2-5a$
18 $3a+5-2-2a$
19 $4a-2b-5a-b$
20 $2x+3x^2-5x-1$
21 $3x+5x^2-2x-3x^2+x$

22 $5a^2+3-7a^2-2$

23 $6-7a+2a$

24 $4x-3y-4x+4y$

25 $ab+2bc-5bc+7ab$

26 $2x-3y+2-8y$

27 $4x^2-7xy+y^2-9x^2$

28 $a-b-3a$

29 $5a+4b-2a-9b$

30 $2x-3-5+8$

12N

Write an answer for each of the following:

1 *a)* The cost of 2 pencils at 5p each.
b) The cost of 3 pencils at 5p each.
c) The cost of n pencils at 5p each.

2 *a)* The cost of n pencils at 5p each.
b) The cost of n pencils at 6p each.
c) The cost of n pencils at m pence each.

3 *a)* The number of bars of chocolate you can get for 60p if each costs 3p.
b) The number of bars of chocolate you can get for 60p if each costs 5p.
c) The number of bars of chocolate you can get for 60p if each costs m pence.

4 *a)* The number of bars of chocolate you can get for q pence if each costs 4p.
b) The number of bars of chocolate you can get for q pence if each costs m pence.
c) The number of bars of chocolate you can get for £1 if each costs m pence.
d) The number of bars of chocolate you can get for £A if each costs m pence.

5 *a)* The total of all the ages of seven girls if their average age is 13 yrs 1 month.
b) The total of all the ages of n girls if their average age is Y years.

6 *a)* The average speed of a car if it covers 106 km in 2 hours.
b) The average speed of a car if it covers d km in t hours.

7 The amount of money in pence each of n girls receive if £X is divided equally among them.

8 The length of a path in metres if it is made up of a slabs each b cms long.

9 The change from £1 after buying n things each costing 2p.

10 The remaining length of string if n pieces each 2 cm long are cut from a piece y metres long.

12O

If $a = 5$, $b = 2$, $c = 0$, $d = 3$ find the value of the following:

1 $a+b+c$ **6** $a+2b-3d$ **11** $\frac{d}{b}$

2 abc **7** $a-(b+d)$ **12** a^2

3 $a+b-d$ **8** $2a-(b+c)$ **13** b^3

4 $d-a$ **9** $d-(a-b)$ **14** $(2a)^2$

5 $2b-a$ **10** $\frac{a}{b}$ **15** $2a^2$

12P

If $a = 1$, $b = -2$, $c = 4$ and $d = 0$, find the value of the following:

1 $a+b+c$ **6** $b+c+d$ **11** b^2

2 $a-b$ **7** $d-2b$ **12** a^3

3 $b-c$ **8** abc **13** $2c^2$

4 $2a+b$ **9** bcd **14** $\frac{a}{c}$

5 $2a-c$ **10** c^2 **15** $\frac{c}{b}$

12Q

If $a = -1$, $b = 3$, $c = -2$, and $d = 5$, find the value of the following:

1 $a+b$ **6** $a+b+c$ **11** ac

2 $b+c+d$ **7** $2d-b$ **12** $2b^2$

3 $b-c$ **8** $2b-d$ **13** c^3

4 $d-a$ **9** bd **14** $\frac{d}{b}$

5 $c-a$ **10** abd **15** $\frac{c}{a}$

12R

If $p = -3$, $q = 2$, $r = 4$ and $s = -6$, find the value of the following:

1 $p+q$ **2** $q+r+s$ **3** $q-p$ **4** $p+s$ **5** $2q-r$

6 $2p-s$ **7** $p+q+r+s$ **8** pq **9** ps **10** $3q^2$

11 p^3 **12** qp^2 **13** $\frac{q}{r}$ **14** $\frac{r}{p}$ **15** $\frac{p}{s}$

12S

Perform the following addition sums.

1 $\begin{array}{r} 3a+2b \\ 2a-b \\ \hline \end{array}$ **2** $\begin{array}{r} 5a+b \\ a-4b \\ \hline \end{array}$ **3** $\begin{array}{r} 2a-6b \\ a-3b \\ \hline \end{array}$

4 $\begin{array}{r} 2a-2b \\ 5a+5b \\ \hline \end{array}$ **5** $\begin{array}{r} 4a-3b \\ -2a-5b \\ \hline \end{array}$

6–10 Set down the same pairs of expressions and subtract instead of adding.

12T

Add together the following pairs of expressions:

1 $\begin{array}{r} a^2+3a \\ 2a^2-a \\ \hline \end{array}$ **2** $\begin{array}{r} 2a+5 \\ -4a-3 \\ \hline \end{array}$ **3** $\begin{array}{r} 2ab-3bc \\ ab-6bc \\ \hline \end{array}$

4 $\begin{array}{r} a^2+3a-2 \\ 2a^2-5a-1 \\ \hline \end{array}$ **5** $\begin{array}{r} 2a^2-5ab-b^2 \\ 4a^2-2ab+4b^2 \\ \hline \end{array}$

6–10 Set down the same expressions and subtract instead of adding.

12U

Multiply together the following pairs of numbers:

1	$+2, -3$	**7**	$-3, +8$	**13**	$-6, -2$	**19**	$-9, -1$
2	$-5, -1$	**8**	$-2, +2$	**14**	$-7, +3$	**20**	$+10, -3$
3	$-4, +2$	**9**	$-8, +1$	**15**	$-3, -4$		
4	$+6, -3$	**10**	$-4, -5$	**16**	$-5, +5$		
5	$+2, +7$	**11**	$-4, +5$	**17**	$-7, +8$		
6	$-9, -2$	**12**	$+4, -5$	**18**	$+2, -5$		

12V

Find the value of the following:

1 $(-2)\times(+7)$

2 $(-3)\times(-5)$

3 $(-5)\times(-6)$

4 $(+8)\times(-6)$

5 $(-10)\times(-1)$

6 $(-9)\times(+3)$

7 $(-4)\times(+9)$

8 $(+3)\times(-8)$

9 $(+11)\times(-4)$

10 $(-7)\times(-4)$

11 $(-1)\times(+1)$

12 $(-3)\times(+2)\times(-4)$

13 $(+5)\times(-1)\times(+2)$

14 $(-6)\times(-2)\times(+3)$

15 $(-2)\times(-3)\times(-4)$

16 $(-4)\times(-4)\times(+5)$

17 $(+3)\times(+6)\times(-2)$

18 $(-2)\times(-5)\times(-9)$

19 $(+3)\times(-6)\times(-7)$

20 $(-5)\times(+3)\times(+2)$

12W

Find the value of the following:

1 $(+3)\div(+2)$

2 $(-6)\div(-2)$

3 $(+4)\div(+8)$

4 $(-5)\div(+2)$

5 $(+5)\div(-2)$

6 $(-5)\div(-2)$

7 $(+3)\div(-9)$

8 $(-9)\div(+3)$

9 $(+12)\div(-3)$

10 $(-12)\div(-4)$

11 Divide $(+24)$ by $(+3)$

12 Divide (-24) by (-3)

13 Divide $(+24)$ by (-3)

14 Divide $(+14)$ by (-7)

15 Divide (-10) by (-20)

16 Divide (-8) by $(+6)$

17 Divide $(+6)$ by (-9)

18 Divide (-6) by (-12)

19 Divide $(+8)$ by (-12)

20 Divide (-1) by $(+3)$

12X

Find the value of each of the following:

1 $\dfrac{(-2)\times(+4)}{(+2)}$

2 $(-3)^2\times(-5)$

3 $\dfrac{(+4)\times(-3)}{(-2)}$

4 $(+8)^2\div(-16)$

5 $\dfrac{(-1)\times(-2)}{(-3)}$

6 $(+7)\div(-2)$

7 $(+2)^2\times(-5)$

8 $\dfrac{(+6)^2}{(-9)}$

9 $\dfrac{(-4)^2\times(-3)}{6}$

10 $\dfrac{(-2)\times(+5)\times(-6)}{-15}$

12Y

If $a = -1$, $b = 3$, $c = 4$, $d = 2$ find the value of the following:

1 $b+c$

2 $bd+cd$

3 $d(b+c)$

4 $b+a$

5 $bc+ac$

6 $c(b+a)$

7 $c+(b+a)$

8 $(c+b)+a$

9 $c-d$

10 $b+(c-d)$

11 $b-(c-d)$

12 $b-c-d$

13 $b-c+d$

14 $b-c$

15 $a(b-c)$

12Z

Simplify the following:

1 $2a+3b+(a+5b)$

2 $2a+3b-(a+5b)$

3 $x^2+2x+(x^2-x)$

4 $x^2+2x-(x^2-x)$

5 $x-3y+(2x+y)$

6 $x-3y-(2x+y)$

7 $3a+(7b-2a)-3b$

8 $4a-(2b-2a+5b)$

9 $3x-6-(4-x)$

10 $3a-8b-(a+2b)$

11 $4-5a-(3a-2)$

12 $10x+3y-(6x-4)$

13 $10x+3y-2(3x-2)$

14 $6a-8b-(5+3b)$

15 $2(3a-4b)-(5a+3b)$

16 $6x^2+xy-3xy+5y^2$

17 $x-(3x^2+2x)-x^2$

18 $2(a+b)-a-b$

19 $5(2x+y)-2y$

20 $3(a^2+a-2)+10$

3 Topology

13A Topological Equivalence

1 Are the following pairs of figures topologically equivalent?

2 Each of these figures has two 3-nodes and two 4-nodes. Pick out pairs which are topologically equivalent.

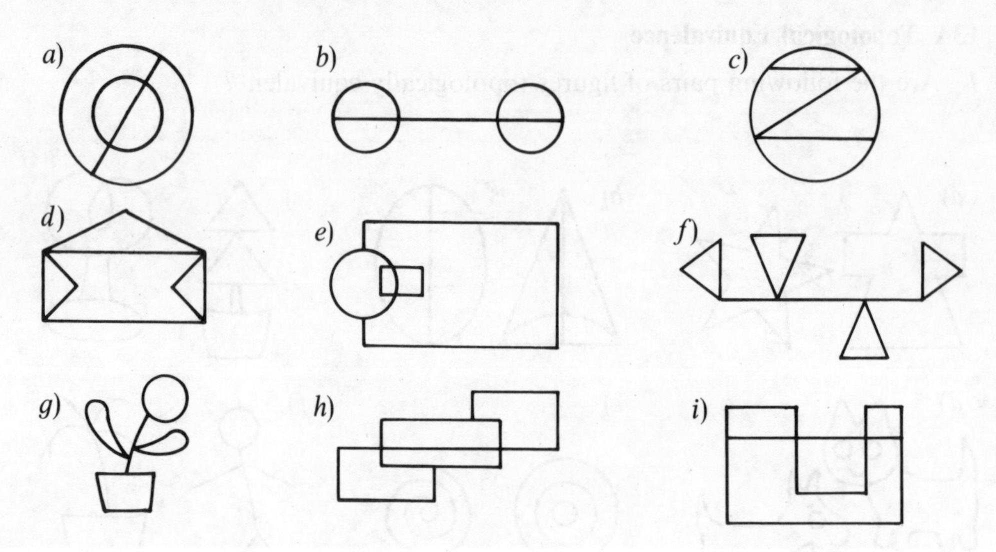

3 State the number and kind of nodes in the following figures:

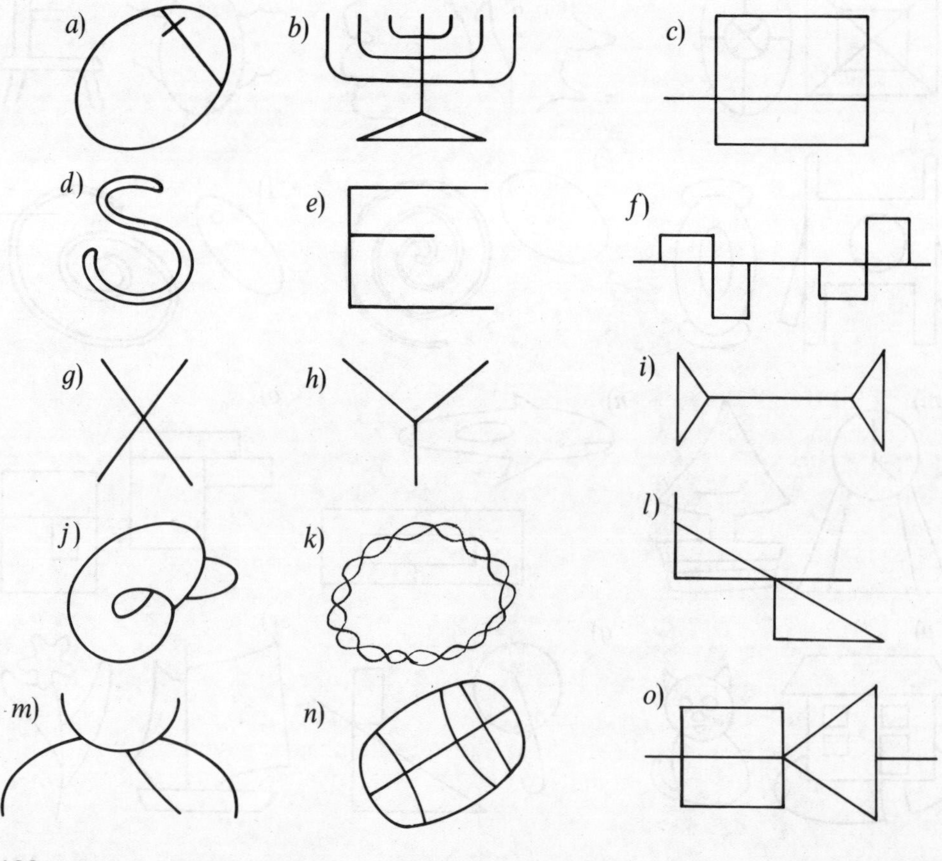

4 The map shows 4 towns on the south coast of Britain and 8 inland towns, and how they are linked by 5 different railway lines.

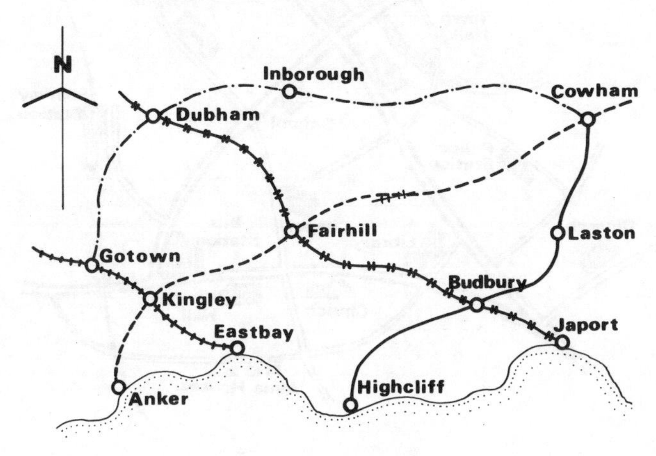

Scale: 1 cm = 4 km

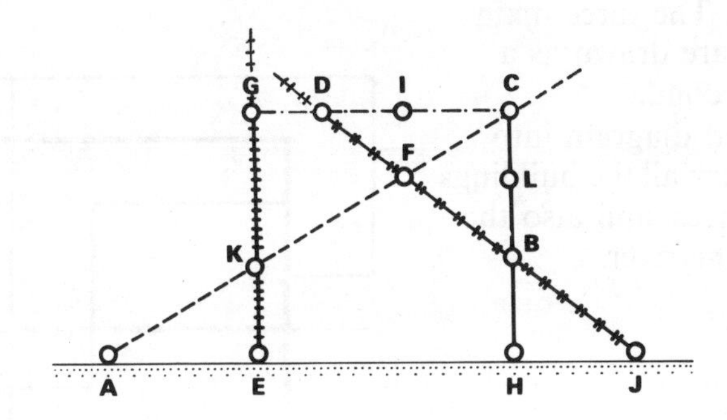

This diagram shows topologically how the 12 towns are linked by the railway lines.

Answer the following questions and state which diagram you look at to find each answer:

a) How far is Dubham from Gotown by rail?
b) If all trains stop at each of the stations through which they pass, how many times would you have to change trains in travelling from Eastbay to Cowham?
c) Which town is due north of Fairhill?
d) How many stations do you pass through in travelling from Inborough to Highcliff?
e) Which two stations is Laston between?
f) It is mid-September and the setting sun is in the engine driver's eyes; which line is he travelling along?
g) If you live at Inborough, which coastal resort can you reach most easily?
h) If you live at Fairhill, which is your nearest coastal resort?
i) How many of these 12 stations are at the end of the journey for some of the trains?

5 The diagram shows some of the roads in a small town and a few buildings.

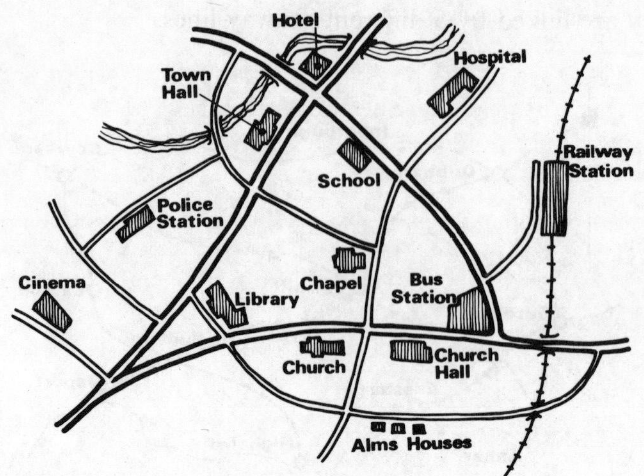

The second diagram is topologically equivalent to the first, showing the same network of roads. The three main roads in the first are drawn as a rectangle in the second.

Copy the second diagram into your book and mark all the buildings in appropriate places, and also the railway line and the river.

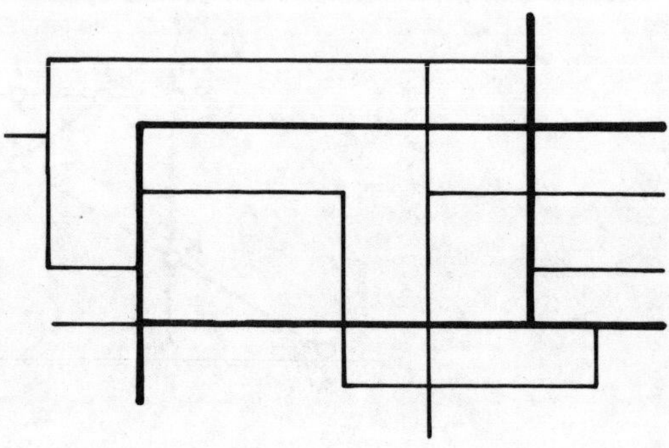

6 Do the following properties remain invariant under a topological transformation? Answer YES or NO.

a) Length of lines
b) Order of points in a line
c) Size of angles
d) Number and type of nodes
e) Area
f) Number of loops
g) 'Inside' or 'outside' a given loop
h) Shape
i) Traversability
j) Number of sides
k) Number of angles
l) Number of diagonals

Answer also the following questions (YES or NO):

m) Do parallel lines remain parallel?
n) Do congruent triangles remain congruent?

o) Do similar triangles remain similar?
p) Do perpendicular lines remain perpendicular?
q) Do regular figures remain regular?
r) If *A*, *B* and *C* are points on a line and *A* is on the right and *C* is on the left of *B*, is this still true after the transformation?
s) If *A*, *B* and *C* are points on a line and *A* lies between *B* and *C*, is this still true after the transformation?

If you understand the meanings of 'translation', 'rotation', 'reflection' or 'enlargement' answer some or all of the following questions:

7 Repeat question 6 for a translation.

8 Repeat question 6 for a rotation.

9 Repeat question 6 for a reflection.

10 Repeat question 6 for an enlargement.

11 Draw figures to the specifications given in the following table. If this is not possible in any particular case, say 'not possible'.

	1-nodes	3-nodes	4-nodes	5-nodes	6-nodes
a)	2	2	–	–	–
b)	–	–	2	–	–
c)	2	–	1	–	1
d)	3	3	–	–	–
e)	1	2	4	1	–
f)	3	–	2	–	–
g)	1	1	1	1	1
h)	2	1	1	1	1
i)	–	2	1	1	–
j)	–	2	1	2	–
k)	4	–	1	–	–
l)	1	–	2	–	–
m)	–	1	1	–	1
n)	–	–	3	–	1
o)	–	1	1	1	–

What can you deduce from your results?

Draw some more diagrams of your own and count the nodes to find out if this is always true.

12 *a*) Print the numbers 1 to 9 in any way you wish.
b) Arrange together in groups numbers which are topologically equivalent.
c) Make a list of the number and type of nodes in each number.
d) Group together numbers with identical arrangements of nodes.
e) Compare the grouping in *b*) and *d*). What do you deduce?

13B Traversability

1 Are the following figures traversable?
Answer YES or NO. If NO, state the number of 'journeys' required to traverse the figure.

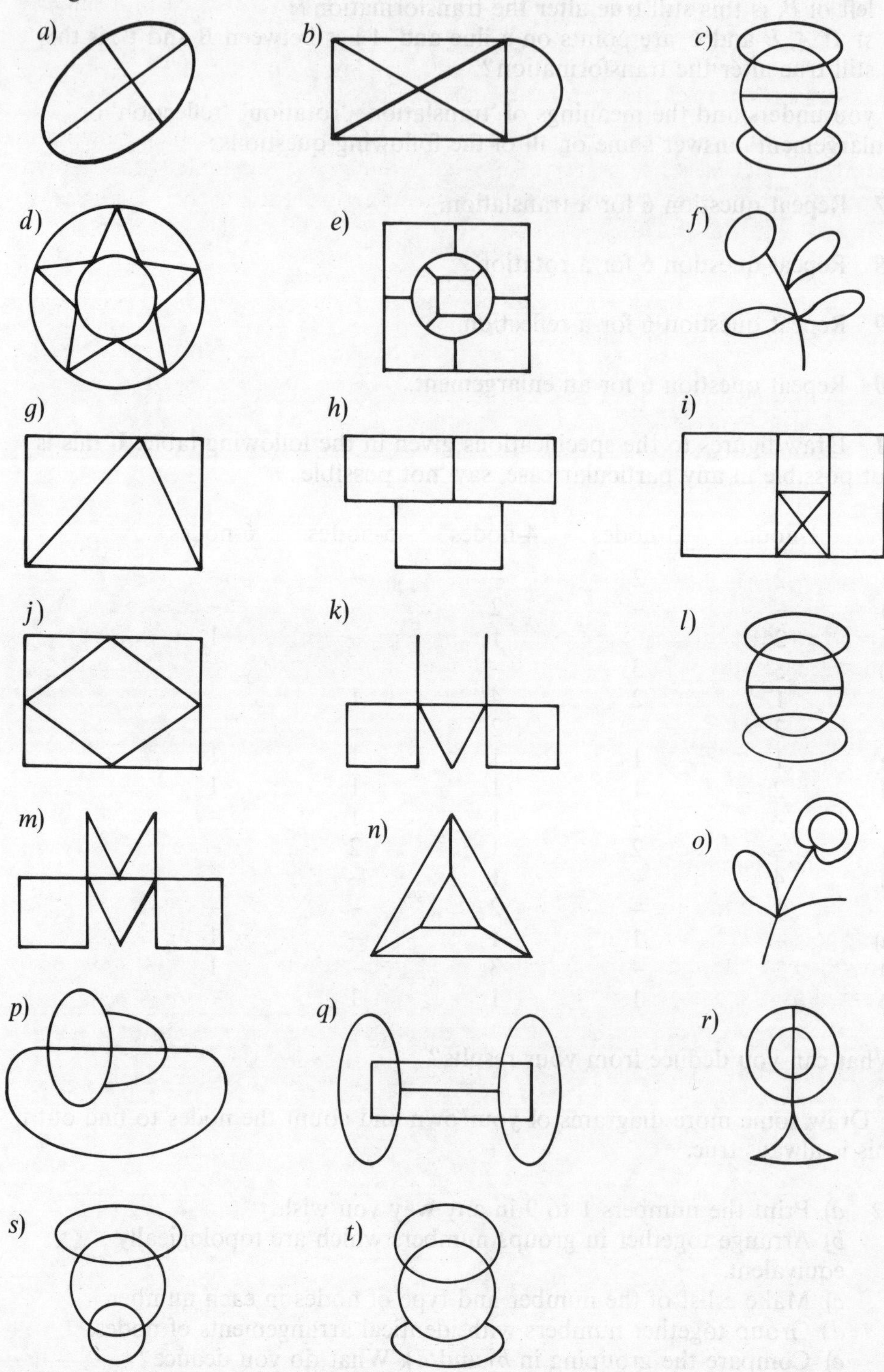

2 Draw up a table showing the number and type of nodes in each of the 20 figures in question 1, and the number of journeys required to traverse it.

3 Some of the figures in question 1 could be traversed in a single journey starting at any point of the figure. Others can only be traversed in a single journey if the journey starts at a specified point, e.g. at a 3-node. Some of the figures could only be traversed in two or more journeys and even then it was necessary to start each journey at a specified point.

Add a column to the table in question 2 (or make a new table) showing the point at which it is necessary to start each journey. Your table should now show the number and type of nodes for each figure, the number of journeys required and the starting point.

Note Do not letter the figures. In cases where a starting point must be specified, describe it as 'a 3-node', etc. In cases where any starting point will do, say 'any starting point'.

∗ *4* *a*) In questions 1 to 3, in cases where a special starting point is required, what do you notice about the finishing point?
b) In cases where no special starting point is required, what do you notice about the finishing point?

∗ **5** Using the tables in questions 2 and 3 and your answers to question 4, can you now draw up a set of rules governing traversability?

∗ *6* Draw 20 figures of your own design. Make tables similar to those in questions 2 and 3. Do your figures confirm the rules you found in question 5?

If not, ask your teacher to check the rules you found and if necessary correct them.

13C The Four Colour Problem

Colour the following networks using four colours or less, but not more than four colours. Each region must be in one colour, and no two regions bounded by the same arc must have the same colour. The outside must be treated as a separate region.

Hint It is better to mark the regions with the numbers 1 to 4, lightly in pencil, until all the snags have been removed. Use colour later.

a)

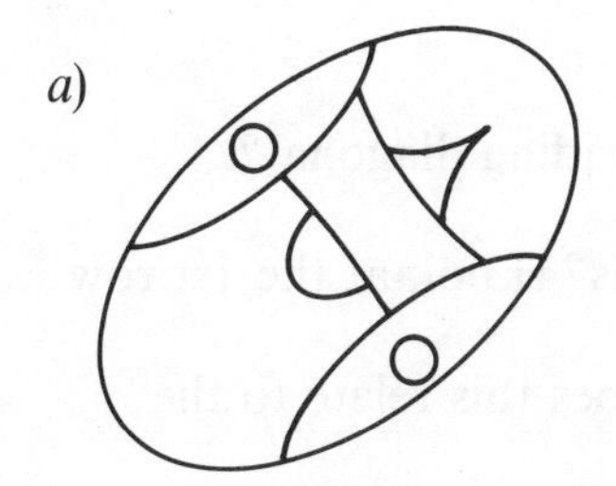

b)

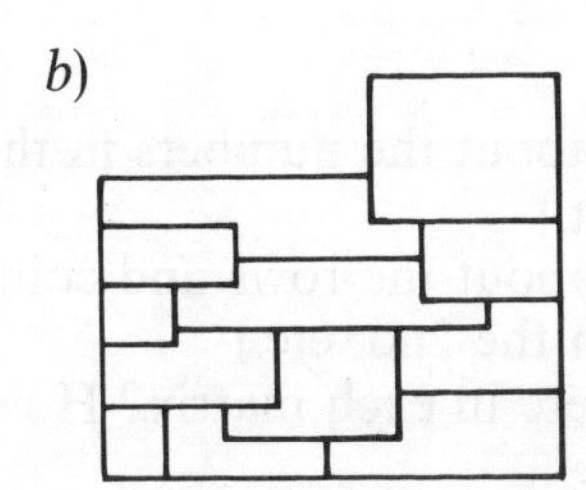

c)

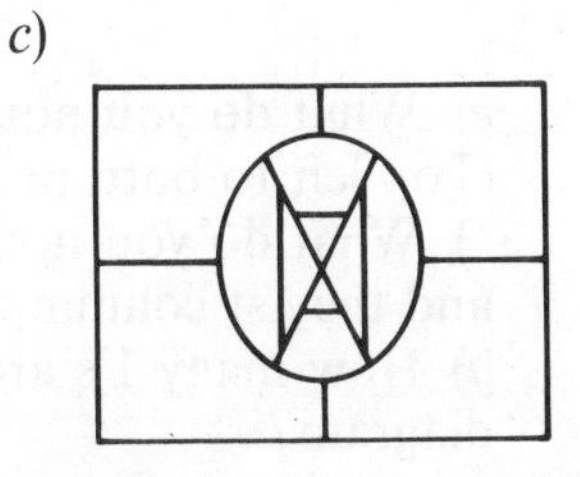

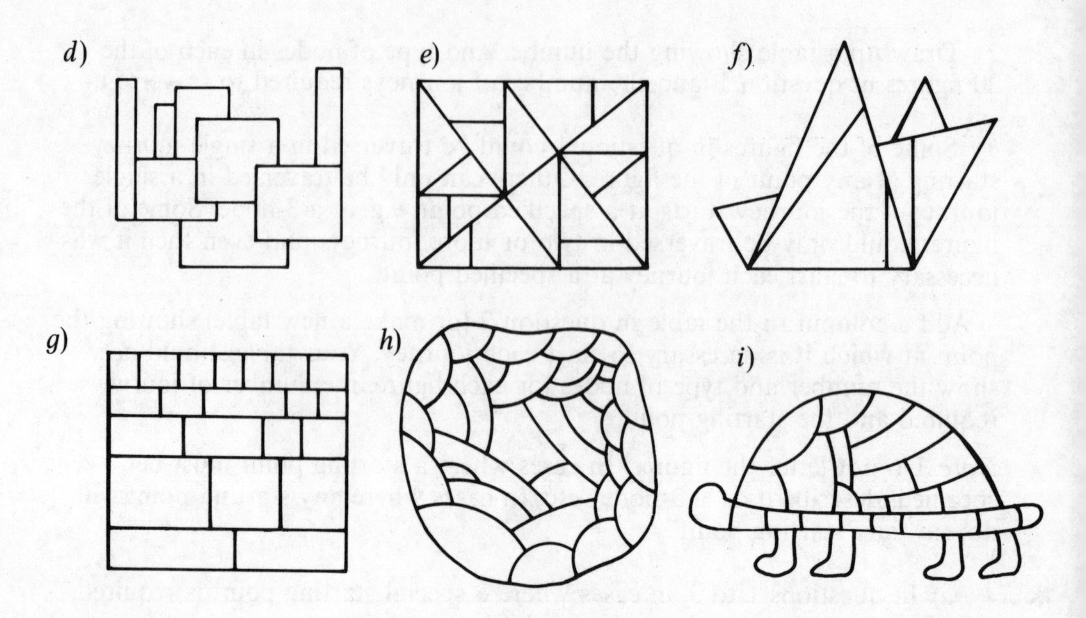

13D Route Matrices

1 Write down the direct route matrix for each of the following networks.

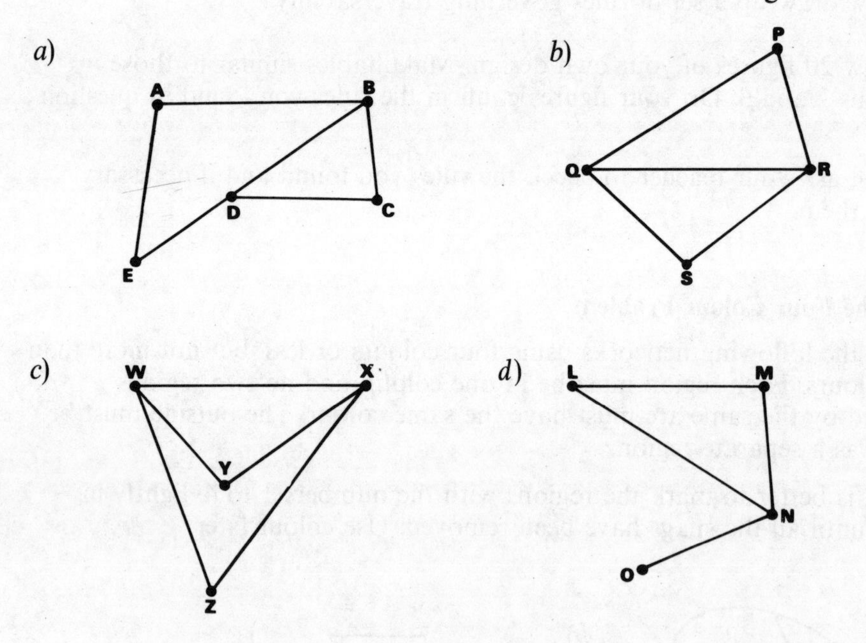

e) What do you notice about the numbers in the leading diagonal? (Top left to bottom right.)
f) What do you notice about the rows and columns? (Look at the 1st row and the 1st column, then the 2nd, etc.)
g) How many 1's are there in each matrix? How does this relate to the diagram?

2 Write down the direct route matrix for each of the following networks.

a)

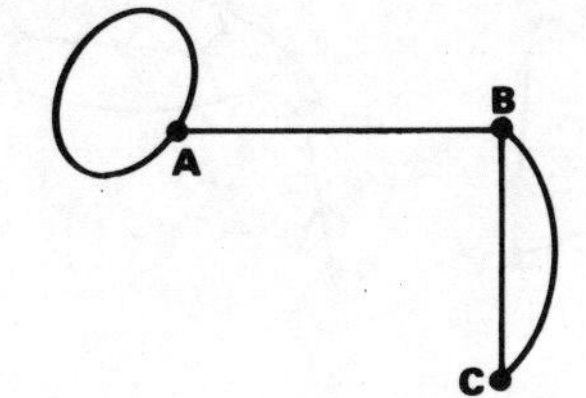

b)

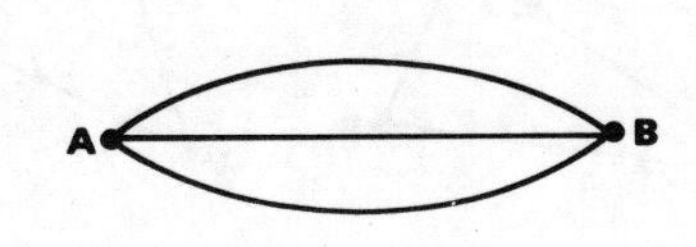

c)

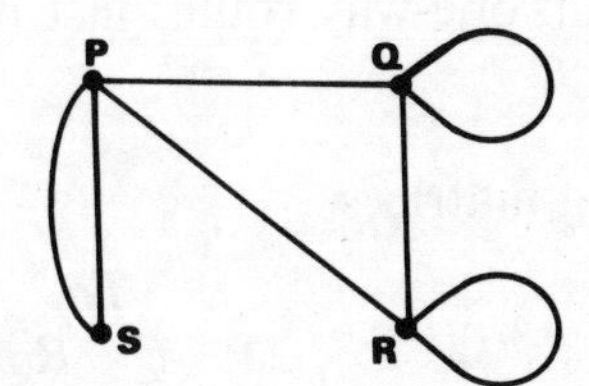

d)

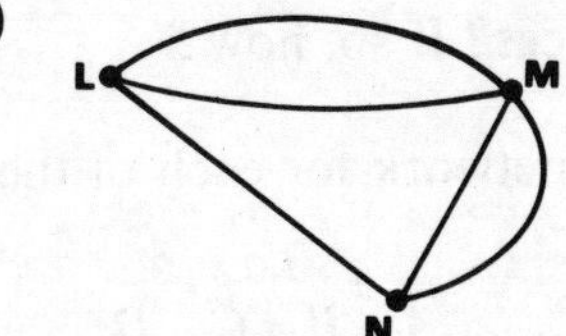

e) Why are there other numbers beside 0 in the leading diagonals?
f) Why are all the numbers not 0's and 1's?

3 Draw networks for each of the following direct route matrices.

a)

	A	*B*	*C*	*D*
A	0	1	1	0
B	1	0	1	1
C	1	1	0	1
D	0	1	1	0

b)

	P	*Q*	*R*
P	0	1	1
Q	1	0	1
R	1	1	0

c)

	L	*M*	*N*
L	2	1	1
M	1	0	2
N	1	2	0

d)

	W	*X*	*Y*	*Z*
W	0	2	0	1
X	2	0	1	0
Y	0	1	2	1
Z	1	0	1	2

e)

	E	*F*	*G*	*H*	*I*
E	0	0	2	0	1
F	0	2	0	1	0
G	2	0	0	3	1
H	0	1	3	0	0
I	1	0	1	0	2

4 Write down the matrices for these networks which contain both one-way routes and two-way routes. Insert 'From' and 'To' as in question 5.

a)

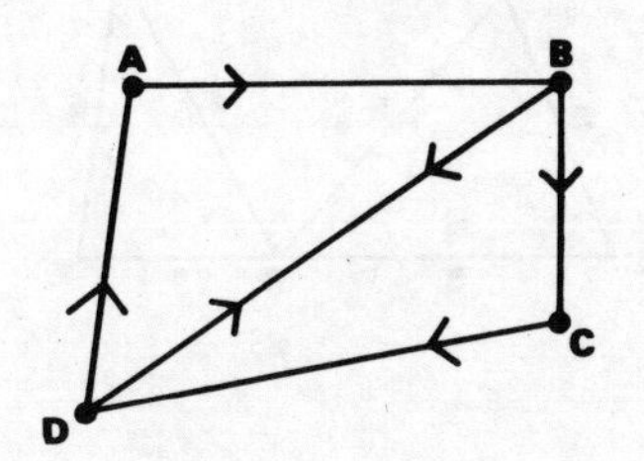

b)

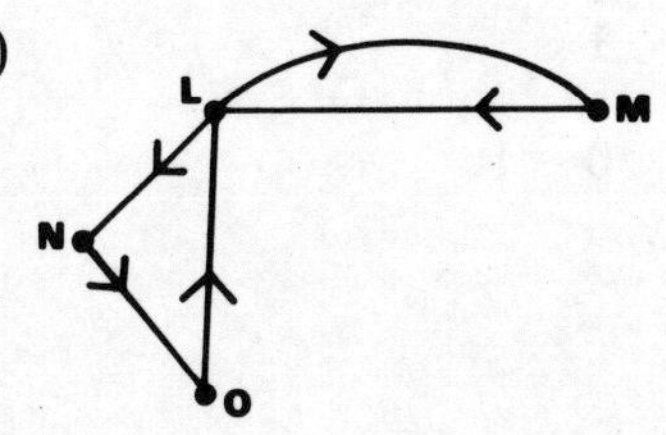

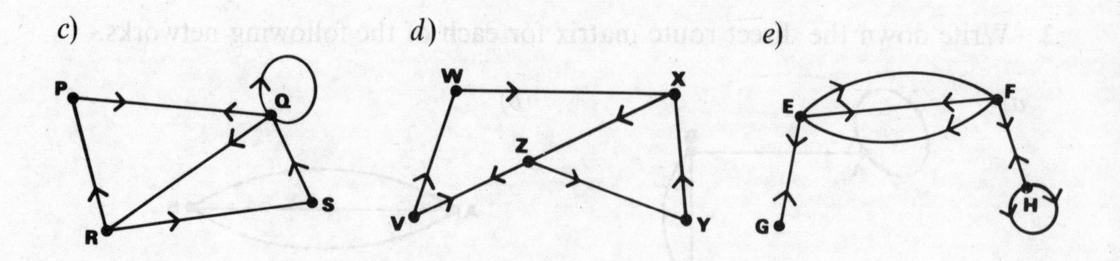

f) Can you tell that these networks contain one-way routes just by looking at the matrices? If so, how?

5 Draw a network for each of the following matrices.

a)

		To			
		A	*B*	*C*	*D*
	A	1	0	1	0
From	*B*	0	0	0	1
	C	1	1	0	1
	D	1	0	0	0

b)

		To			
		P	*Q*	*R*	*S*
	P	0	2	0	1
From	*Q*	2	0	1	1
	R	1	0	0	1
	S	1	1	0	0

c)

		To				
		L	*M*	*N*	*O*	*P*
	L	0	1	1	1	0
	M	0	0	2	0	1
From	*N*	1	2	0	1	1
	O	1	1	0	0	1
	P	0	0	1	1	1

d)

		To		
		X	*Y*	*Z*
	X	1	2	1
From	*Y*	2	0	0
	Z	0	1	1

e)

		To				
		V	*W*	*X*	*Y*	*Z*
	V	0	1	0	0	1
	W	0	1	0	2	0
From	*X*	1	0	0	1	1
	Y	0	2	0	2	1
	Z	0	0	1	0	0

6 The matrix below, known as an 'incidence' matrix, relates to network *a*). Complete the matrix and write down similar matrices for networks *b*), *c*), *d*) and *e*).

	A	*B*	*C*	*D*	*E*
1	1	1	0	0	0
2	1		1		
3	0	0	1		
4					
5		0	0	1	
6					
7					

a)

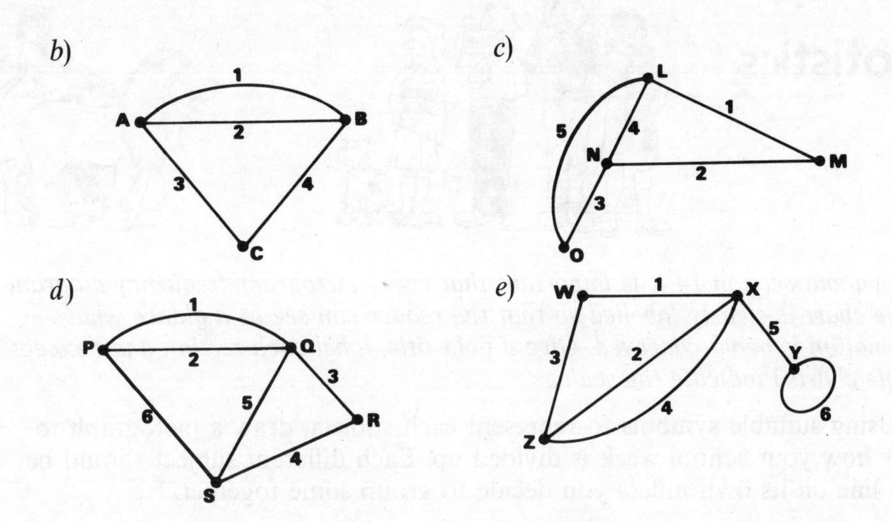

7 In the matrices in question 6, what do the numbers in the columns tell you about the nodes? Give examples.

* *8* Can you think of any practical purpose to which the type of topology taught in this chapter can be put?

* *9* In the light of your answer to question 8, do you think it worth while studying topology?

14 Statistics

14A

Throughout section 14 it is important that every pictogram, frequency diagram or pie chart is clearly labelled so that the reader can see at a glance what information is being conveyed. Give a bold title, label each section and (except for pie charts) indicate the scale.

1 Using suitable symbols to represent each subject, draw a pictograph to show how your school week is divided up. Each different subject should be on a line on its own unless you decide to group some together.

2 Choose eight popular television programmes and ask everyone in your class to name the two which they like best out of your list of eight. Draw a pictogram to illustrate your findings. When you choose your eight programmes, remember that you will have to design a symbol to represent each one. You may find it necessary to let one symbol stand for, say, three votes.

3 Note the colours of all the cars in the school car park. If necessary group together some shades. Draw a bar chart to show your findings.

4 Count up all the different pets that you and the members of your class keep. Draw a bar chart to show these totals. Draw also a pictogram.

5 Draw a bar chart to show the number of brothers and sisters each member of your class has. Arrange the columns in order: no brothers or sisters; one brother or sister; two; three, etc.

6 A sweet-pea grower selected 40 stalks and noted how many flowers there were on each. These were his findings:

2	3	4	5	3	2	1	3
4	3	1	7	5	4	4	6
6	5	4	3	5	4	1	5
5	7	3	5	4	7	5	6
5	5	3	6	2	2	6	5

Find the frequency with which each of these numbers of flowers occurs and draw a frequency diagram.

7 List the heights of all the members of your class. Group them in ranges of 3 to 5 cm, or whatever is reasonable, to give between 8 and 12 groups altogether. Draw a frequency diagram to show your findings.

8 Get every one in your class to throw a die 20 times and record the number of times each of 1, 2, 3, 4, 5 and 6 is thrown. Draw a frequency diagram to show the results. What would you expect it to look like?

9 Record the results when each member of your class tosses two coins 20 times. Draw a diagram to show the frequency of two heads, one head and one tail, and two tails. What should this diagram look like?

10 Ask each member of your class how long it takes them to travel to school. Group these times in suitable intervals such as 0–10 minutes; 10–20 minutes, etc. Draw a frequency graph with about 8 or more columns showing your results.

14B

1 60 pupils in the final year were leaving school. 16 were going to universities, 14 to colleges of education, 11 into engineering, 6 into nursing, 4 into the services and 9 into secretarial training. Show this information in a pie chart.

2 30 pupils in a class were asked to name their favourite game. 6 picked hockey, 14 football, 7 basketball and 3 tennis. Show this information in a pie chart.

3 Find out how each of the members of your class travels to school and draw a pie chart showing the results. Approximate small angles where necessary.

4 A boy classified the 45 books he had on his shelves and drew a pie chart showing the results. The angles he constructed were:

Animal stories	88°
School stories	48°
Adventure books	120°
Travel books	40°
Others	64°

How many books of each type did he have?

5 Two girls kept a record of the 480 grades that they had been given in a year and drew pie charts to show the results. The angles that they measured in their charts were as follows:

Anne			Kate		
	A	48°		A	30°
	A−	81°		A−	66°
	B+	99°		B+	84°
	B	63°		B	126°
	B−	42°		B−	33°
	C	27°		C	21°

Find out how many times each girl actually had each grade and draw one frequency diagram to show both sets of results.

6

'If you ever, ever, ever, ever,
If you ever, ever, ever meet a whale,
You must never, never, never, never,
You must never, never, never pull its tail.
For if you ever, ever, ever, ever,
If you ever, ever, ever pull its tail,
You will never, never, never, never,
You will never, never meet another whale.'

a) Count the number of words in this rhyme.
b) Count the number of times each of the following words occur: ever, never, you, if, other words (lumped together).
c) Draw a pie chart to illustrate the distribution of words in the rhyme.

14C

1 *a*) Find the arithmetic mean of these ten numbers:
1 2 5 6 6 9 11 14 15 17
b) Find also the modal value of the same set of numbers.

2 Find the mean and the mode of this set of numbers:
8 10 2 3 5 6 1 3 11 15 6 9 9 7 6 12 3 5 6 11

3 Find the mean and the mode of these numbers:
19 15 27 21 38 21 35 15 21 29

4 Find the median and mean of these sets of numbers:

a) 2 5 6 6 8 10 15 15 19
b) 3 4 7 8 8 21 25 26 30
c) 1 2 2 4 6 6 9 11 15 16
d) 3 6 7 7 10 11 14 15 15 20
e) 4 4 6 7 8 11 12 12 14 15

5 Find the median of these sets of numbers:

a) 5 4 7 9 21 3 17 5 11
b) 14 3 6 3 18 10 7 3 1
c) 15 4 1 6 25 13 16 8 10 12
d) 22 17 25 13 14 21 24 16
e) 39 16 25 27 32 36 18 22

6 Ten pupils were asked to draw a line 4 cm long without measuring it. The actual lengths were then found to the nearest 0·5 cm. The results were: 2 had drawn 4·5 cm, 3 had drawn 4·0 cm, 4 had drawn 3·5 cm and 1, 3·0 cm. Find the average length of the lines drawn.

7 A class of 20 were asked to draw an angle of 60° without using a protractor. The angles measured to the nearest 5° were as follows:

2 were 45°	4 were 55°	4 were 65°
6 were 50°	3 were 60°	1 was 75°

Find the average angle drawn.

8 Using your frequency table for question 6 in exercise 14A find the average number of flowers per stalk on the sweet peas.

9 Two schoolboys who were keen on statistics stood beside a level crossing and wrote down the number of people in each of the cars going over the crossing.

23 cars had only 1 person
15 cars had 2 people
3 cars had 3 people
6 cars had 4 people
3 cars had 6 people

Find the average number of people per car.

10 Find the mean of these 25 numbers by using the 'guessed mean' method. Start by using 20 as a working mean

22	15	16	19	24
24	18	23	22	22
17	18	25	15	23
20	16	20	19	18
21	25	21	17	20

11 Find the average age of forty schoolchildren whose actual ages are as follows. Use the 'guessed mean' method and give your answer to the nearest month.

12 y 3 m	12 y 1 m	11 y 11 m	12 y 5 m	12 y 2 m
11 y 8 m	12 y 1 m	11 y 11 m	11 y 10 m	12 y 2 m
12 y 0 m	11 y 8 m	12 y 4 m	11 y 8 m	12 y 0 m
11 y 5 m	11 y 10 m	12 y 4 m	11 y 6 m	12 y 5 m
12 y 4 m	12 y 3 m	11 y 6 m	12 y 4 m	11 y 7 m
12 y 2 m	12 y 1 m	12 y 4 m	12 y 5 m	11 y 10 m
11 y 7 m	11 y 9 m	11 y 7 m	12 y 1 m	11 y 7 m
11 y 8 m	12 y 0 m	11 y 7 m	11 y 9 m	11 y 8 m

12 Find the average age of your class in the same way as in question 11.

13 The table shows the heights of 30 girls, measured to the nearest cm. Find the average height, giving your answer also to the nearest cm.

130	150	128	136	148	128	140	139	139	138
135	142	141	145	132	150	148	150	142	140
140	131	133	139	135	132	133	145	142	145

14 Find the average height of all the members of your class.

15 A volunteer working for a Consumers' Advice Bureau recorded the price of a bottle of sherry of a certain brand in 30 different shops and supermarkets. These are the prices she recorded:

£1·02	99p	£1.14	£1.45	£1.02	£1.15
£1.05	£1.15	99p	£1.11	£1.05	£1.14
£1.11	£1.15	£1.15	95p	£1.05	£1.35
95p	£1.10	£1.15	97p	97p	£1.20
98p	£1.02	£1.10	97p	95p	97p

Find the average price.

16 The average age of a family of six children was 5 years 8 months. What was the total of all their ages?

17 *a*) The average sum of money that each of five boys had in his pocket was 46p. How much did they have altogether?
b) John had the most: he had £1. When he left them, what was the average amount that each of the other four boys had?

18 *a*) A driver found that the average distances he had covered per day for six days was 240 km. How far had he driven altogether?
b) On the last day he had only driven 120 km. What was the average per day for the first five days?

14D

1 Find the median and the mode of these ordered sets of numbers:

a) 10 12 12 15 15 15 18 24 24
b) 2 4 4 9 10 12 12 12 17 17
c) 25 27 27 27 30 31 31 33 39

2 Find the median and the mean of these sets of numbers:

a) 2 7 1 10 11 17 7 2 9 5 14
b) 24 30 21 22 17 19 30 19 27
c) 105 110 121 97 102 112 100 107
d) 92 82 84 98 85 80 78 92 97 82

3 Find the mean of these numbers using the 'guessed mean' method:

49 47 45 51 52 48
54 50 52 55 49 46
52 51 47 54 47 55
52 49 47 54 51 49
48 46 50 46 49 55

4 On certain match boxes the average contents are quoted as 74 matches per box. The number of matches in each of 20 boxes was counted. The results were as follows:

No. of matches per box	71	72	73	74	75	76
No. of boxes	4	5	0	1	6	4

Find the average number of matches per box in this sample of 20 boxes.

5

Pocket money	25p	50p	75p	£1.00	£1.25	£1.50	£1.75	£2.00
No. of boys	3	5	3	15	12	9	5	8

The table shows the result of a survey carried out to find how much pocket money each of 60 boys had. Find the average amount per boy.

6 In five cricket matches an opening batsman scored the following runs in his ten innings:

34 51 10 35 72 49 55 69 16 42

Find the average number of runs he scored per innings.

7 *a*) A fruiterer opened twenty cases, each holding 60 oranges. He found the numbers of bad ones in each box were as follows:

No. of bad oranges	0	1	2	3	4	5	6	9
No. of boxes	2	0	3	3	5	4	2	1

Find the average number of bad oranges per box.

b) He bought another twenty cases from a different importer, and this time the numbers of bad oranges in each case were as under:

No. of bad oranges	0	1	2	3	4	5	8
No. of boxes	3	2	6	1	3	3	2

Find the average number of bad oranges per box.

Which importer should he deal with in future? Discuss.

8 Write down an ordered set of five integers whose mean is 4 and whose median is 4.

9 The heights of 30 conifer seedlings were measured after eighteen months of growth. The heights are given to the nearest cm:

Height	16	17	18	19	20	21	22	23
No. of seedlings	4	3	7	8	3	2	1	2

Find the mean height of this batch of seedlings.

10 The heights in cm of another 30 seedlings were noted separately:

20	23	22	23	20	19
22	20	17	17	21	21
17	17	19	18	20	22
18	17	20	23	18	17
23	21	20	24	17	17

Find the mean height of this batch of seedlings, having first grouped the data as in the previous question.

14E

1 Find a set of seven integers whose median is 5 and whose mean is 5.

2 Write down an ordered set of five integers with a mean of 7, a mode of 10 and a median of 10.

3 The average age of a group of 10 children is 11 years 10 months. Find the total of all their ages.

4 The heights of 30 children, measured to the nearest cm, were noted as follows:

152	149	149	146	153	145
147	151	149	147	152	148
149	151	150	145	150	151
155	146	153	151	146	154
152	148	150	154	147	147

Using the 'guessed mean' method, find the average height of these children.

5 The heights of another 30 children, in the same age group as the 30 in the previous question, were noted. Their heights were taken to the nearest even number of cm:

Height in cm	144	146	148	150	152	154	156
No. of children	2	4	8	9	5	1	1

Find the mean height of these children.

6 In a cable car at a holiday camp the number of people travelling in each of 50 cable cars was recorded:

No. of people	1	2	3	4	5	6
No. of cars	20	11	6	7	4	2

Find the average number of people per cable car.

7 The number of passengers in each of 50 vehicles going on to a ferry was observed:

No. of passengers	1	2	3	4	5	6
No. of vehicles	3	8	10	25	3	1

Find the average number of passengers per vehicle.

8 The table shows the number of people staying each night in a seaside hotel over a period of one month:

No. of people	10	12	15	20	21	22	23	24
No. of nights	3	3	1	2	1	2	8	11

Find the average number staying per night during that month.

9 The number of goals scored by a hockey team at each match over a season of 20 matches was as follows:

No. of goals	0	1	2	3	4	6
No. of matches	4	2	4	6	3	1

Find the average number of goals per match.

10 The number of goals scored by the different opposing teams in the same 20 matches was also noted:

No. of goals	0	1	2	3	4	5	6
No. of matches	2	3	4	5	3	1	2

Find the average number of goals scored against the team in the previous question.

∗ *11* Your answer to number 10 should be larger than your answer to number 9. Does this mean the team in number 9 lost more matches than they won? Discuss.

15 Elementary Vectors, Translations

15A Vectors

Definitions A vector such as $\binom{3}{2}$ can be drawn anywhere in the co-ordinate plane. It is then called a *free* vector. If it starts from the origin, it is called a *position* vector. If it is drawn in a specified line, it is called a *line* vector.

For questions 1 to 3, draw a pair of co-ordinate axes in the middle of a sheet of graph paper. Choose a suitable scale, allowing at least ± 8 on either axis.

1 Draw the following free vectors, giving three positions of each and using all four quadrants.

a) $\begin{pmatrix} 5 \\ 3 \end{pmatrix}$ *b)* $\begin{pmatrix} 2 \\ 1 \end{pmatrix}$ *c)* $\begin{pmatrix} -3 \\ 2 \end{pmatrix}$ *d)* $\begin{pmatrix} 4 \\ -2 \end{pmatrix}$

e) $\begin{pmatrix} -5 \\ -3 \end{pmatrix}$ *f)* $\begin{pmatrix} 6 \\ 10 \end{pmatrix}$ *g)* $\begin{pmatrix} 12 \\ 4 \end{pmatrix}$

2 In each part of this question you are given a line and a line vector. Draw the line in pencil and mark at least two positions of the vector in ink or colour.

a) x-axis $\begin{pmatrix} 4 \\ 0 \end{pmatrix}$ *b)* y-axis $\begin{pmatrix} 0 \\ 2 \end{pmatrix}$ *c)* $x+y=0$ $\begin{pmatrix} 4 \\ -4 \end{pmatrix}$

d) $x+y=6$ $\begin{pmatrix} 4 \\ -4 \end{pmatrix}$ *e)* $y=2x+4$ $\begin{pmatrix} 3 \\ 6 \end{pmatrix}$ *f)* $3y+x=6$ $\begin{pmatrix} -9 \\ 3 \end{pmatrix}$

3 In each part of this question you are given a position vector. Draw this vector, label it, and give the co-ordinates of its tip, as in the sketch.

a) $\begin{pmatrix} 3 \\ 5 \end{pmatrix}$ *b)* $\begin{pmatrix} -4 \\ 2 \end{pmatrix}$ *c)* $\begin{pmatrix} 4 \\ -2 \end{pmatrix}$ *d)* $\begin{pmatrix} -4 \\ -2 \end{pmatrix}$

e) $\begin{pmatrix} 0 \\ -2 \end{pmatrix}$ *f)* $\begin{pmatrix} 3 \\ 0 \end{pmatrix}$ *g)* $\begin{pmatrix} -1 \\ 5 \end{pmatrix}$ *h)* $\begin{pmatrix} -3 \\ 4 \end{pmatrix}$

$\begin{pmatrix} 2 \\ 3 \end{pmatrix}$ (2,3)

What do you deduce about the co-ordinates of a point and the position vector of that point?

4 Do not use graph paper for this question.
Sketch freehand a set of axes and draw the following free vectors. Mark them as shown.

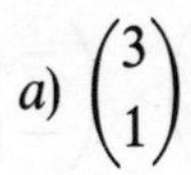

a) $\begin{pmatrix} 3 \\ 1 \end{pmatrix}$ *b)* $\begin{pmatrix} -2 \\ 2 \end{pmatrix}$ *c)* $\begin{pmatrix} -1 \\ -2 \end{pmatrix}$

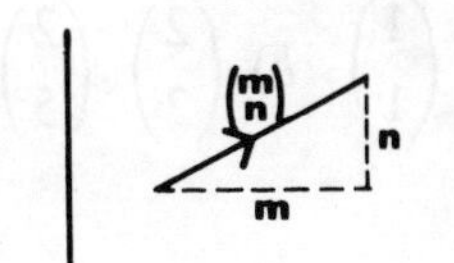

d) $\begin{pmatrix} -4 \\ 3 \end{pmatrix}$ *e)* $\begin{pmatrix} 3 \\ -4 \end{pmatrix}$ *f)* $\begin{pmatrix} -3 \\ 4 \end{pmatrix}$

5 Repeat question 4, but treat the vectors as position vectors. This time mark in also the co-ordinates of the tip of the vector.

6 Without drawing the vectors, state the co-ordinates of the tips of the following position vectors.

a) $\begin{pmatrix} 5 \\ 1 \end{pmatrix}$ *b)* $\begin{pmatrix} 2 \\ -3 \end{pmatrix}$ *c)* $\begin{pmatrix} -4 \\ 7 \end{pmatrix}$ *d)* $\begin{pmatrix} -3 \\ -11 \end{pmatrix}$

e) $\begin{pmatrix} a \\ b \end{pmatrix}$ *f)* $\begin{pmatrix} 2a \\ -3b \end{pmatrix}$ *g)* $\begin{pmatrix} a+b \\ c+d \end{pmatrix}$

7 In each part of this question you are given a free vector and the co-ordinates of its starting point. State the co-ordinates of its tip. Do not draw the vectors.

a) $(5, 2)$ $\begin{pmatrix} 3 \\ 4 \end{pmatrix}$ *b)* $(3, 1)$ $\begin{pmatrix} 7 \\ 4 \end{pmatrix}$ *c)* $(-2, -1)$ $\begin{pmatrix} 4 \\ 5 \end{pmatrix}$

d) $(3, 6)$ $\begin{pmatrix} -5 \\ -2 \end{pmatrix}$ *e)* $(-4, -3)$ $\begin{pmatrix} 2 \\ 8 \end{pmatrix}$ *f)* $(-6, -9)$ $\begin{pmatrix} -3 \\ -4 \end{pmatrix}$

8 In each part of this question you are given a free vector and the co-ordinates of its tip. State the co-ordinates of its starting point. Do not draw the vectors.

a) $\begin{pmatrix} 3 \\ 2 \end{pmatrix}$ $(5, 4)$ *b)* $\begin{pmatrix} 6 \\ 9 \end{pmatrix}$ $(2, 2)$ *c)* $\begin{pmatrix} -3 \\ -4 \end{pmatrix}$ $(-4, -1)$

d) $\begin{pmatrix} 5 \\ -2 \end{pmatrix}$ $(-3, 8)$ *e)* $\begin{pmatrix} -4 \\ -1 \end{pmatrix}$ $(-6, -2)$ *f)* $\begin{pmatrix} 4 \\ -4 \end{pmatrix}$ $(-4, 4)$

9 Give the position vectors of the following points.

a) $(2, 4)$ *b)* $(-1, 3)$ *c)* $(-2, -1)$

d) $(5, -2)$ *e)* $(-5, 0)$ *f)* $(3, -4)$

10 In each part of this question you are given the position vectors of the vertices *A* and *B* of a square *ABCD*. State the position vectors of *C* and *D* (two possible answers for each).

a) $\begin{pmatrix}4\\1\end{pmatrix}\begin{pmatrix}1\\1\end{pmatrix}$ *b)* $\begin{pmatrix}2\\2\end{pmatrix}\begin{pmatrix}2\\5\end{pmatrix}$ *c)* $\begin{pmatrix}3\\1\end{pmatrix}\begin{pmatrix}1\\3\end{pmatrix}$ *d)* $\begin{pmatrix}2\\5\end{pmatrix}\begin{pmatrix}5\\1\end{pmatrix}$ *e)* $\begin{pmatrix}3\\-4\end{pmatrix}\begin{pmatrix}-1\\-1\end{pmatrix}$

15B

1 Add the following vectors:

a) $\begin{pmatrix}3\\5\end{pmatrix}+\begin{pmatrix}2\\1\end{pmatrix}$ *b)* $\begin{pmatrix}4\\2\end{pmatrix}+\begin{pmatrix}3\\1\end{pmatrix}$ *c)* $\begin{pmatrix}5\\1\end{pmatrix}+\begin{pmatrix}2\\-4\end{pmatrix}$

d) $\begin{pmatrix}-1\\-1\end{pmatrix}+\begin{pmatrix}2\\3\end{pmatrix}$ *e)* $\begin{pmatrix}-2\\4\end{pmatrix}+\begin{pmatrix}4\\-2\end{pmatrix}$ *f)* $\begin{pmatrix}3\\-1\end{pmatrix}+\begin{pmatrix}-3\\1\end{pmatrix}$

g) $\begin{pmatrix}-2\\-1\end{pmatrix}+\begin{pmatrix}-4\\3\end{pmatrix}$ *h)* $\begin{pmatrix}-5\\1\end{pmatrix}+\begin{pmatrix}-2\\2\end{pmatrix}$ *i)* $\begin{pmatrix}-3\\6\end{pmatrix}+\begin{pmatrix}-2\\-8\end{pmatrix}$

2 Subtract the following vectors:

a) $\begin{pmatrix}3\\2\end{pmatrix}-\begin{pmatrix}1\\1\end{pmatrix}$ *b)* $\begin{pmatrix}4\\3\end{pmatrix}-\begin{pmatrix}2\\2\end{pmatrix}$ *c)* $\begin{pmatrix}6\\6\end{pmatrix}-\begin{pmatrix}1\\5\end{pmatrix}$

d) $\begin{pmatrix}6\\5\end{pmatrix}-\begin{pmatrix}6\\2\end{pmatrix}$ *e)* $\begin{pmatrix}3\\2\end{pmatrix}-\begin{pmatrix}-2\\-1\end{pmatrix}$ *f)* $\begin{pmatrix}5\\1\end{pmatrix}-\begin{pmatrix}4\\6\end{pmatrix}$

g) $\begin{pmatrix}2\\3\end{pmatrix}-\begin{pmatrix}-1\\-4\end{pmatrix}$ *h)* $\begin{pmatrix}-2\\3\end{pmatrix}-\begin{pmatrix}-3\\2\end{pmatrix}$ *i)* $\begin{pmatrix}-1\\-2\end{pmatrix}-\begin{pmatrix}-1\\-2\end{pmatrix}$

3 Draw graphs on squared paper to illustrate question 1.

4 Draw graphs on squared paper to illustrate question 2.

5 A fly runs on a window pane from one corner *A* to the opposite corner *B*. Taking the horizontal and vertical edges of the pane through *A* as the *X* and *Y* axes, its path is represented by the following vectors, the unit being centimetres.

$$\begin{pmatrix}9\\3\end{pmatrix}\begin{pmatrix}12\\6\end{pmatrix}\begin{pmatrix}9\\-3\end{pmatrix}\begin{pmatrix}15\\12\end{pmatrix}\begin{pmatrix}12\\15\end{pmatrix}\begin{pmatrix}9\\-6\end{pmatrix}\begin{pmatrix}6\\18\end{pmatrix}$$

a) State a vector representing $\vec{AB}$.
b) What are the dimensions of the window pane?

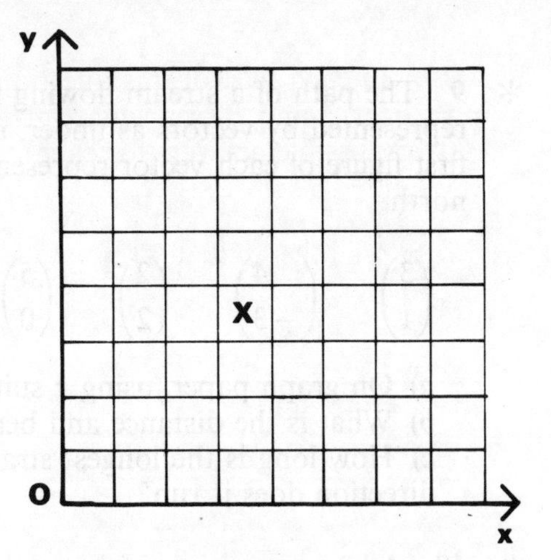

6 (For chess enthusiasts.)
A chess piece stands in the square marked X which can be represented by the position vector $\binom{4}{4}$

a) If the piece is a queen, give vectors for every possible move. (There are 27 of them.)
Do the same for *b*) a rook
c) a knight *d*) a bishop
e) a pawn *f*) a king.

* **7** At a seaside resort a zig-zag path is constructed up the face of a steep cliff which runs due east and west. The sections of the path can be represented by vectors, the unit being metres and the positive directions being west and up. (For instance the vector $\binom{50}{4}$ would mean that the path ran 50 metres west and 4 metres up.)
These are the vectors:

$$\begin{pmatrix}40\\6\end{pmatrix} \begin{pmatrix}-30\\4\end{pmatrix} \begin{pmatrix}50\\4\end{pmatrix} \begin{pmatrix}0\\8\end{pmatrix} \begin{pmatrix}-60\\5\end{pmatrix} \begin{pmatrix}-30\\4\end{pmatrix} \begin{pmatrix}40\\6\end{pmatrix} \begin{pmatrix}-20\\2\end{pmatrix} \begin{pmatrix}50\\7\end{pmatrix}$$

a) What is the height of the cliff?
b) What is the approximate total length of the path?
c) What would you expect to find at the section $\binom{0}{8}$?
d) How far is the end of the path east (or west) of the starting point?

* **8** If the cliff face in question 7 was not vertical but sloped inwards to the north, a third figure could be added to the vectors in question 7 giving the distance gone north in each section of the path. Here are the modified vectors:

$$\begin{pmatrix}40\\6\\1\end{pmatrix} \quad \begin{pmatrix}-30\\4\\1\end{pmatrix} \quad \begin{pmatrix}50\\4\\2\end{pmatrix} \quad \begin{pmatrix}0\\8\\8\end{pmatrix} \quad \begin{pmatrix}-60\\5\\0\end{pmatrix}$$

$$\begin{pmatrix}-30\\4\\1\end{pmatrix} \begin{pmatrix}40\\6\\2\end{pmatrix} \quad \begin{pmatrix}-20\\2\\2\end{pmatrix} \quad \begin{pmatrix}50\\7\\2\end{pmatrix}$$

a) How far does the cliff face slope inwards?
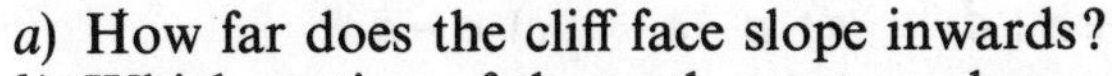
b) Which section of the path most needs a guard rail?
c) Excluding the fourth, which section of the path climbs most steeply?
d) Give fresh details about what you would expect to find in the section $\begin{pmatrix}0\\8\\8\end{pmatrix}$

* 9 The path of a stream flowing from a village A to a village B can be represented by vectors as under, the unit being a length of 100 metres. The first figure of each vector represents a distance east, and the second a distance north.

$$\begin{pmatrix}3\\1\end{pmatrix} \quad \begin{pmatrix}4\\-3\end{pmatrix} \quad \begin{pmatrix}2\\2\end{pmatrix} \quad \begin{pmatrix}5\\0\end{pmatrix} \quad \begin{pmatrix}2\\-1\end{pmatrix} \quad \begin{pmatrix}3\\1\end{pmatrix} \quad \begin{pmatrix}1\\0\end{pmatrix}$$

a) On graph paper, using a suitable scale, draw a sketch of the stream.
b) What is the distance and bearing of B from A?
c) How long is the longest straight stretch of the stream and in what direction does it run?

* 10 A hare runs in a zig-zag path across a meadow from P to Q, its path being represented by vectors, where the unit is a length of 5 metres.

$$\begin{pmatrix}3\\4\end{pmatrix} \quad \begin{pmatrix}8\\2\end{pmatrix} \quad \begin{pmatrix}6\\-3\end{pmatrix} \quad \begin{pmatrix}4\\4\end{pmatrix} \quad \begin{pmatrix}5\\-1\end{pmatrix} \quad \begin{pmatrix}9\\7\end{pmatrix} \quad \begin{pmatrix}8\\3\end{pmatrix} \quad \begin{pmatrix}7\\9\end{pmatrix}$$

A greyhound starts from P 5 seconds later than the hare and running on a straight course intercepts it at Q.

a) Give a vector to represent the path of the greyhound.
b) On graph paper, using a suitable scale, draw a sketch showing the paths of the hare and the greyhound.
c) From your graph, estimate the total distance run by the hare and by the greyhound.
d) If the hare runs at 20 metres per second, how fast must the greyhound run?

Note Use the same scale on both axes or your measured distances will be wrong.

15C

For this exercise,

$$\boldsymbol{a} = \begin{pmatrix}1\\2\end{pmatrix} \quad \boldsymbol{b} = \begin{pmatrix}3\\4\end{pmatrix} \quad \boldsymbol{c} = \begin{pmatrix}-2\\1\end{pmatrix} \quad \boldsymbol{d} = \begin{pmatrix}1\\-3\end{pmatrix} \quad \boldsymbol{e} = \begin{pmatrix}-2\\-4\end{pmatrix}$$

$$\boldsymbol{f} = \begin{pmatrix}3\\6\end{pmatrix} \quad \boldsymbol{g} = \begin{pmatrix}-4\\2\end{pmatrix} \quad \boldsymbol{h} = \begin{pmatrix}-6\\3\end{pmatrix} \quad \boldsymbol{m} = \begin{pmatrix}1\\-4\end{pmatrix} \quad \boldsymbol{n} = \begin{pmatrix}-6\\12\end{pmatrix}$$

1 Give the values of

a) $\boldsymbol{a}+\boldsymbol{b}$ *b)* $\boldsymbol{a}-\boldsymbol{c}$ *c)* $\boldsymbol{b}+\boldsymbol{c}+\boldsymbol{d}$ *d)* $\boldsymbol{a}-\boldsymbol{e}$ *e)* $\boldsymbol{e}-\boldsymbol{a}$.

2 Give the values of

a) $2\boldsymbol{a}$ (i.e. $\boldsymbol{a}+\boldsymbol{a}$) *b)* $3\boldsymbol{a}$ (i.e. $\boldsymbol{a}+\boldsymbol{a}+\boldsymbol{a}$) *c)* $4\boldsymbol{a}$ *d)* $7\boldsymbol{a}$ *e)* $11\boldsymbol{a}$
f) $\frac{1}{2}\boldsymbol{e}$ *g)* $\frac{1}{2}\boldsymbol{g}$ *h)* $\frac{1}{3}\boldsymbol{f}$ *i)* $^{-}\frac{1}{3}\boldsymbol{h}$ *j)* $\frac{2}{3}\boldsymbol{n}$

3 Give the values of

a) $2\boldsymbol{a}+3\boldsymbol{b}$ *b*) $\boldsymbol{a}-2\boldsymbol{b}$ *c*) $\boldsymbol{b}+2\boldsymbol{c}$ *d*) $\boldsymbol{c}-2\boldsymbol{d}+3\boldsymbol{e}$ *e*) $2\boldsymbol{a}-\boldsymbol{e}+3\boldsymbol{c}$

f) $\boldsymbol{d}-2\boldsymbol{b}$ *g*) $3\boldsymbol{c}+\boldsymbol{a}$ *h*) $2\boldsymbol{b}-\boldsymbol{d}$ *i*) $2\boldsymbol{a}-2\boldsymbol{c}+\boldsymbol{e}$ *j*) $4\boldsymbol{a}-4\boldsymbol{c}+2\boldsymbol{e}$

k) $\frac{1}{2}\boldsymbol{g}+\frac{1}{3}\boldsymbol{f}$ *l*) $\boldsymbol{a}-\frac{1}{2}\boldsymbol{e}$ *m*) $\frac{1}{2}\boldsymbol{a}+\frac{1}{2}\boldsymbol{b}$ *n*) $\frac{1}{2}(\boldsymbol{a}+\boldsymbol{b})$ *o*) $\frac{1}{3}\boldsymbol{f}-\frac{1}{2}\boldsymbol{b}-\frac{1}{2}\boldsymbol{m}$

4 If $\boldsymbol{i}=\begin{pmatrix}1\\0\end{pmatrix}$ and $\boldsymbol{j}=\begin{pmatrix}0\\1\end{pmatrix}$ then the vectors $\boldsymbol{i}$ and $\boldsymbol{j}$ are called base vectors.

Then $\boldsymbol{a}=\boldsymbol{i}+2\boldsymbol{j}$ and $\boldsymbol{b}=3\boldsymbol{i}+4\boldsymbol{j}$.

Give the values of $\boldsymbol{c}, \boldsymbol{d}, \boldsymbol{e} \ldots \boldsymbol{n}$ in terms of $\boldsymbol{i}$ and $\boldsymbol{j}$.

5 Give the values of the vectors listed in question 3 in terms of $\boldsymbol{i}$ and $\boldsymbol{j}$.

6 If O is the origin and the point A has the co-ordinates stated, give two expressions for the position vector of the point A.

(Example: A is $(3,2)$ OA is $\begin{pmatrix}3\\2\end{pmatrix}$ or $3\boldsymbol{i}+2\boldsymbol{j}$)

a) $(3,1)$ *b*) $(-2,1)$ *c*) $(4,2)$ *d*) $(5,-3)$ *e*) $(-3,5)$

f) $(-3,-3)$ *g*) $(0,0)$ *h*) $(7,1)$ *i*) $(1,4)$ *j*) $(-2,3)$

7 If $\boldsymbol{p}$ is $\begin{pmatrix}3\\2\end{pmatrix}$ and $\boldsymbol{q}$ is $\begin{pmatrix}6\\4\end{pmatrix}$ then $\boldsymbol{p}$ is parallel to $\boldsymbol{q}$ and twice as long.

If $\boldsymbol{q}$ is $\begin{pmatrix}-6\\-4\end{pmatrix}$ then $\boldsymbol{p}$ and $\boldsymbol{q}$ are still parallel but in the opposite sense.

Pick out any pairs of parallel vectors in the list $\boldsymbol{a}$ to $\boldsymbol{n}$ in question 1. If they are in the opposite sense, say so.

8 State which of the following vectors are parallel: if they are in the opposite sense, say so.

$3\boldsymbol{i}+4\boldsymbol{j}$ $2\boldsymbol{i}+6\boldsymbol{j}$ $\boldsymbol{i}+3\boldsymbol{j}$ $6\boldsymbol{i}+8\boldsymbol{j}$ $-\boldsymbol{i}-3\boldsymbol{j}$ $2\boldsymbol{i}-6\boldsymbol{j}$

$\boldsymbol{i}-3\boldsymbol{j}$ $-\boldsymbol{i}+3\boldsymbol{j}$ $3\boldsymbol{i}-4\boldsymbol{j}$ $9\boldsymbol{i}+12\boldsymbol{j}$

9 The vectors $\binom{r}{s}$ and $\binom{s}{-r}$ are perpendicular. So also are any multiples of these vectors.

Which of the following vectors are perpendicular?

$$\begin{pmatrix}3\\2\end{pmatrix} \begin{pmatrix}2\\-3\end{pmatrix} \begin{pmatrix}-3\\2\end{pmatrix} \begin{pmatrix}-2\\3\end{pmatrix} \begin{pmatrix}-4\\6\end{pmatrix} \begin{pmatrix}4\\-6\end{pmatrix} \begin{pmatrix}4\\6\end{pmatrix} \begin{pmatrix}1\\2\end{pmatrix} \begin{pmatrix}2\\-1\end{pmatrix} \begin{pmatrix}4\\-2\end{pmatrix}$$

10 Which of the following vectors are either parallel or perpendicular?

$2\boldsymbol{i}+3\boldsymbol{j}$ $\boldsymbol{i}-\boldsymbol{j}$ $-\boldsymbol{i}-\boldsymbol{j}$ $\boldsymbol{i}+\boldsymbol{j}$ $3\boldsymbol{i}+2\boldsymbol{j}$ $3\boldsymbol{i}-2\boldsymbol{j}$

$4\boldsymbol{i}+6\boldsymbol{j}$ $3\boldsymbol{i}+3\boldsymbol{j}$ $4\boldsymbol{i}-4\boldsymbol{j}$ $8\boldsymbol{i}+12\boldsymbol{j}$

15D

1 On graph paper plot these eight points:

$A\ (1,0)$ $B\ (3,1)$ $C\ (2,3)$ $D\ (0,2)$
$E\ (-1,1)$ $F\ (-2,-1)$ $G\ (-1,-2)$ $H\ (3,-1)$

Write down the following as column vectors:

$\overline{AB}$ $\overline{BD}$ $\overline{DF}$ $\overline{FG}$ and $\overline{GA}$

Draw these lines on the diagram. What do you notice about the shape which you make? What do you notice about the sum of the column vectors? Must this always be the case if a series of vectors make a closed shape?

2 Using the same points as in the previous question, write down the column vectors $\overline{DA}$ and $\overline{CB}$. What do you notice about them and the lines DA and CB?

Write down the column vectors $\overline{AB}$ and $\overline{DC}$. What do you notice about them? What shape is $ABCD$? Can you find another pair of vectors which are the same as one another?

3 Using the same points, write down the vectors $\overline{GF}$ and $\overline{HD}$. What do you notice about them? What do you notice about the lines GF and HD?

Find another pair of vectors which have a similar property.

4 Draw the following vectors on graph paper using the same scale on both axes.

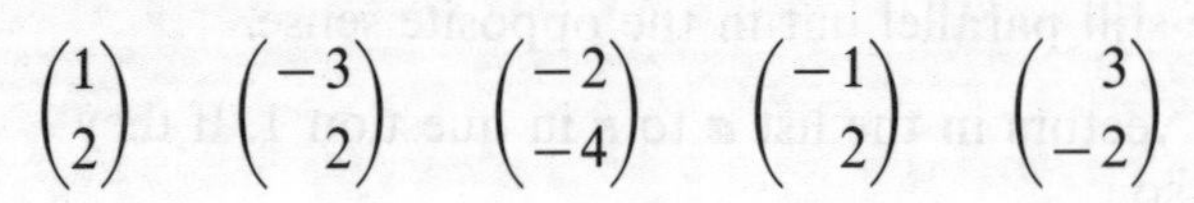

$$\begin{pmatrix}1\\2\end{pmatrix} \quad \begin{pmatrix}-3\\2\end{pmatrix} \quad \begin{pmatrix}-2\\-4\end{pmatrix} \quad \begin{pmatrix}-1\\2\end{pmatrix} \quad \begin{pmatrix}3\\-2\end{pmatrix}$$

Pick out pairs of lines which are equal in length and those which are parallel to one another. Look at their vectors.

5 $A\ (1,0)$ $L(3,1)$ and $P\ (2,4)$ are three of the vertices of a parallelogram. Write down the column vector $\overline{AL}$.
If X is the fourth vertex of parallelogram $ALXP$, write down the vector $\overline{PX}$ and hence the co-ordinates of X.
Write down the column vectors $\overline{AP}$ and $\overline{LX}$.

It is possible to complete the parallelogram in two other ways.
Find these parallelograms; call them $ALPY$ and $AZLP$. Write the co-ordinates of Y and Z.
In each parallelogram write the vectors of one pair of parallel sides.

6 If $\boldsymbol{a} = \begin{pmatrix}-1\\3\end{pmatrix} + \begin{pmatrix}3\\1\end{pmatrix}$ $\boldsymbol{b} = \begin{pmatrix}-1\\-3\end{pmatrix} + \begin{pmatrix}-1\\-1\end{pmatrix}$ $\boldsymbol{c} = \begin{pmatrix}-1\\1\end{pmatrix} + \begin{pmatrix}2\\1\end{pmatrix}$

$\boldsymbol{d} = \begin{pmatrix}-2\\-1\end{pmatrix} + \begin{pmatrix}4\\5\end{pmatrix}$ $\boldsymbol{e} = \begin{pmatrix}0\\-2\end{pmatrix} + \begin{pmatrix}-1\\0\end{pmatrix}$

by drawing these pairs of vectors or otherwise, find $\boldsymbol{a}$, $\boldsymbol{b}$, $\boldsymbol{c}$, $\boldsymbol{d}$ and $\boldsymbol{e}$.
Write down anything you can find about pairs of lines representing the vectors $\boldsymbol{a}$, $\boldsymbol{b}$, $\boldsymbol{c}$, $\boldsymbol{d}$ and $\boldsymbol{e}$.
What do you notice about the directions of the vectors?

7 If $\boldsymbol{a} = \begin{pmatrix} 4 \\ -1 \end{pmatrix}$ $\quad \boldsymbol{b} = \begin{pmatrix} 1 \\ 3 \end{pmatrix}$ $\quad \boldsymbol{c} = \begin{pmatrix} 2 \\ 2 \end{pmatrix}$

without drawing diagrams give the column vectors equivalent to:

a) $-\boldsymbol{a}$, $-\boldsymbol{b}$, $-\boldsymbol{c}$ *b*) $2\boldsymbol{a}$, $3\boldsymbol{b}$, $2(-\boldsymbol{c})$ *c*) $\boldsymbol{a}+3\boldsymbol{b}$, $\boldsymbol{b}+(-\boldsymbol{c})$, $\boldsymbol{c}+2\boldsymbol{a}$

d) $2\boldsymbol{a}+(-\boldsymbol{c})$, $-\boldsymbol{b}+\boldsymbol{c}$, $3\boldsymbol{b}+2(-\boldsymbol{c})$ *e*) $\frac{1}{2}\boldsymbol{c}$, $\frac{1}{2}\boldsymbol{c}+\frac{1}{2}\boldsymbol{b}$, $\frac{1}{2}(\boldsymbol{a}+\boldsymbol{b})$

8 Using $\boldsymbol{a}$, $\boldsymbol{b}$ and $\boldsymbol{c}$ as given in the previous question and without drawing a diagram, find three different possible values for $\boldsymbol{d}$ and $\boldsymbol{e}$ such that if all five vectors were drawn one after the other, they would make a closed shape.
Would the order in which $\boldsymbol{a}$, $\boldsymbol{b}$ and $\boldsymbol{c}$ were drawn make any difference to $\boldsymbol{d}$ and $\boldsymbol{e}$?

9 *ABCD* is a parallelogram. *E* and *F* are the mid-points of *AB* and *DC* respectively.
If $\boldsymbol{a}=\overline{DA}$ and $\boldsymbol{b}=\overline{DF}$ write the following in terms of $\boldsymbol{a}$ and $\boldsymbol{b}$ only:

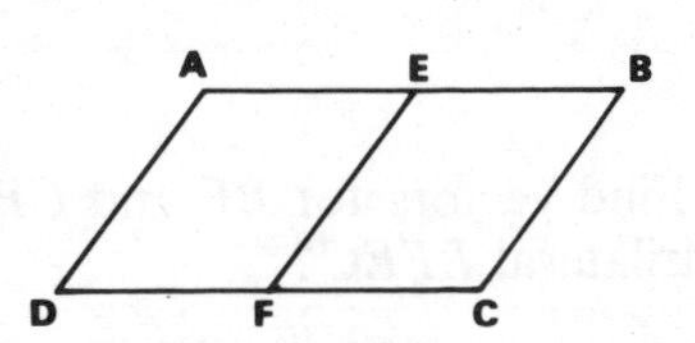

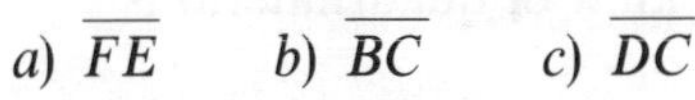

a) $\overline{FE}$ *b*) $\overline{BC}$ *c*) $\overline{DC}$

d) $\overline{BA}$ *e*) $\overline{DE}$ *f*) $\overline{CE}$

10 *ABC* is a triangle with *X* and *Y* the mid-points of *AB* and *BC* respectively.
If $\boldsymbol{a}=\overline{AX}$ and $\boldsymbol{b}=\overline{CY}$ write the following in terms of $\boldsymbol{a}$ and $\boldsymbol{b}$:

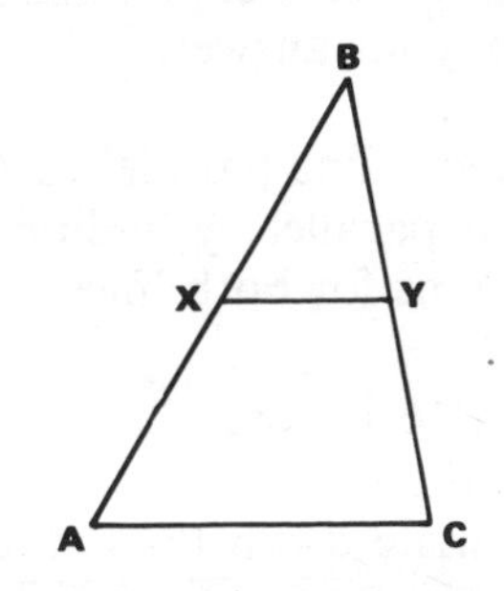

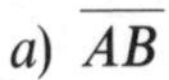

a) $\overline{AB}$ *b*) $\overline{BC}$ *c*) $\overline{AC}$ *d*) $\overline{XY}$

What do the last two answers tell you about the lines *AC* and *XY*?

11 If the three triangles in the diagram are all equilateral and $\boldsymbol{a}=\overline{AE}$ and $\boldsymbol{b}=\overline{CD}$, write the following in terms of $\boldsymbol{a}$ and $\boldsymbol{b}$:

$\overline{EB}$ $\quad \overline{AB}$ $\quad \overline{ED}$ $\quad \overline{AC}$ $\quad \overline{CE}$

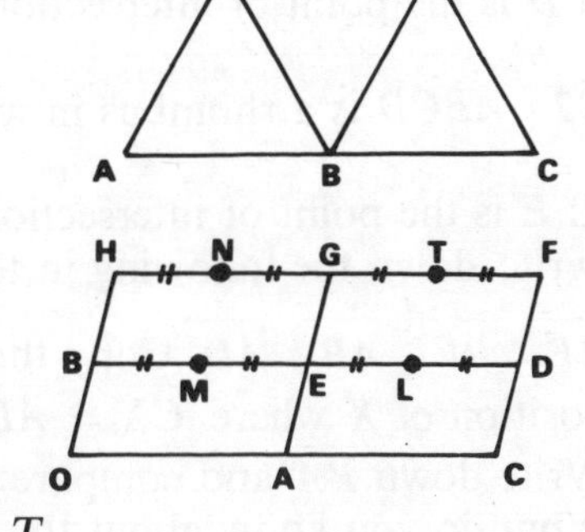

12 The lines *HF*, *BD* and *OC* are parallel, and also the lines *OH*, *AG*, *CF*.
If $\overline{OA}=\boldsymbol{a}$ and $\overline{OB}=\boldsymbol{b}$, give vectors for *OC*, *OD*, *OE*, *OF*, *OG*, *OH*. With *O* as origin give the position vectors for *L*, *M*, *N*, and *T*.

13 Give vectors for BG, EF, AD. What do you deduce?
Give vectors for OD, BF, AB, CH, DG. What do you deduce?

14 Give vectors for LN, TM, NM, TL, ML, TN. What can you say about the quadrilateral $LMNT$?

15 Give vectors for CM, DN. What kind of quadrilateral is $CMND$?

16 $ABCD$ is a regular hexagon centre O. $PQRSTU$ is a similar hexagon with parallel sides, the same centre O and all sides double the length.
If $OA = \boldsymbol{a}$ $OB = \boldsymbol{b}$ find vectors for OC, OD, OE, OF.

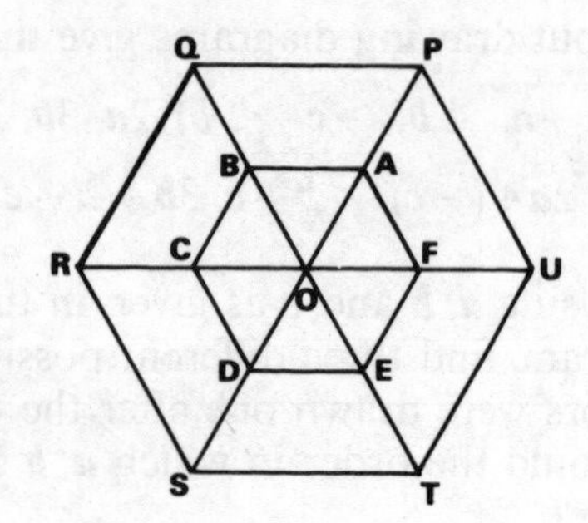

17 Find vectors for BF and CE. What do you deduce about the quadrilateral $BFEC$?

18 Find vector expressions for BU, CT. What kind of quadrilateral is $BUTC$?

19 Find vector expressions for RP, PU, UT, TR. Add these together. Explain your answer.

20 Find a line parallel to RD. Check that it is parallel by finding the vector expressions for both lines.

21 $OACB$ is a parallelogram, $\overline{OA} = \boldsymbol{a}$ and $\overline{OB} = \boldsymbol{b}$.
If M is the mid-point of AC and N is the midpoint of OA, write down the following vectors in terms of $\boldsymbol{a}$ and $\boldsymbol{b}$: $\overline{OC}$ $\overline{ON}$ $\overline{NM}$.
What do you know about OC and NM?
If D is the point of intersection of OC and AB, give the vectors for OD and BD.

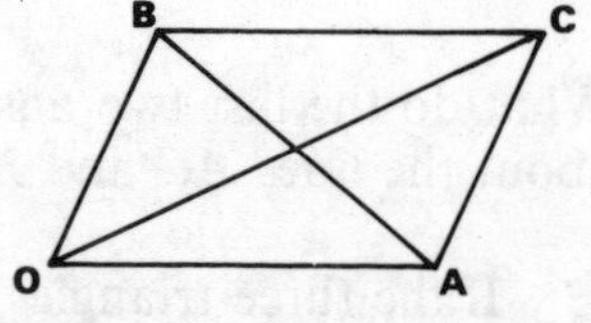

22 $ABCD$ is a rhombus in which $\overline{AC} = \boldsymbol{p}$ and $\overline{BD} = \boldsymbol{q}$.

If E is the point of intersection of the diagonals, write down the following in terms of $\boldsymbol{p}$ and $\boldsymbol{q}$:

$\overline{AE}$ $\overline{BE}$ $\overline{AB}$ $\overline{AD}$. Copy the diagram and mark in the position of X where $\overline{CX} = \overline{AD}$. What shape is $ACXD$?
Write down $\overline{BA}$ and compare it with $\overline{AD}$.
What do you know about the lines AB and AD?

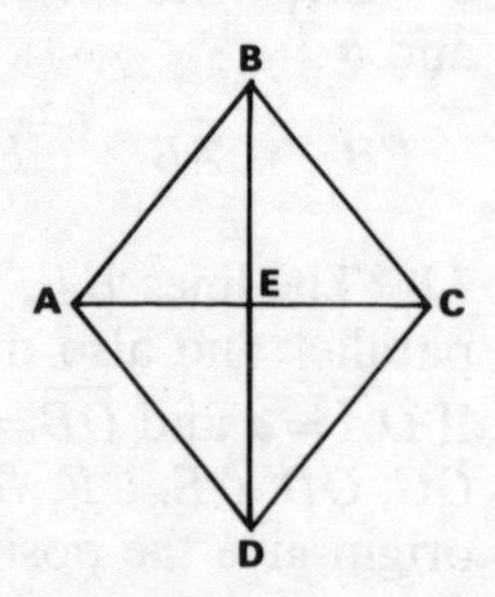

15E Translations

HE SAYS THIS SECTION IS ABOUT TRANSLATIONS.

1 State the images of the following points under the translation $\binom{4}{3}$:

a) $(3,2)$ *b*) $(4,1)$
c) $(-2,4)$ *d*) $(5,3)$
e) $(-2,-4)$ *f*) $(4,-2)$
g) $(2,5)$ *h*) (a,b)
i) (p,q) *j*) $(0,0)$

2 State in each case the translation which maps the following points on to the point $(2,1)$:

a) $(3,1)$ *b*) $(1,3)$ *c*) $(2,2)$ *d*) $(1,1)$
e) $(0,0)$ *f*) $(5,4)$ *g*) $(-3,-2)$ *h*) $(-4,2)$
i) $(4,-7)$ *j*) $(-8,-9)$

3 The point $(3,1)$ could be mapped on to the point $(5,2)$ by the two translations $\binom{1}{1}$ and $\binom{1}{0}$ applied successively: or equally well by the two translations $\binom{4}{3}$ and $\binom{-2}{-2}$. For each of the following pairs of points state two translations which, applied successively, will map the first point on to the second. (There is, of course, more than one possible answer in each case.)

a) $(3,2)$ and $(4,2)$ *b*) $(3,2)$ and $(5,1)$
c) $(-1,4)$ and $(4,-2)$ *d*) $(0,6)$ and $(-3,0)$
e) $(-1,-4)$ and $(-4,2)$ *f*) (a,b) and (c,d)

4 If in the diagram opposite the vector $\boldsymbol{p}$ produces a translation which maps square H on to E, on to which squares will the following be mapped by the given translations?

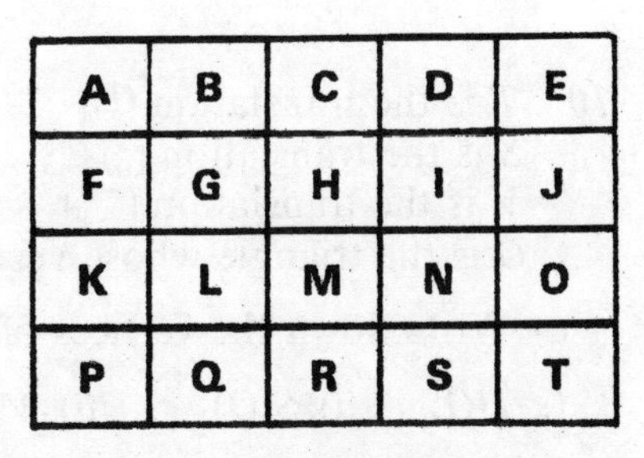

a) $K \rightarrow$ by $\boldsymbol{p}$
b) $F \rightarrow$ by $\boldsymbol{p}$
c) $E \rightarrow$ by $-\boldsymbol{p}$
d) $P \rightarrow$ by $2\boldsymbol{p}$
e) $J \rightarrow$ by $-2\boldsymbol{p}$

Questions 5, 6 and 7 refer to the diagram below.

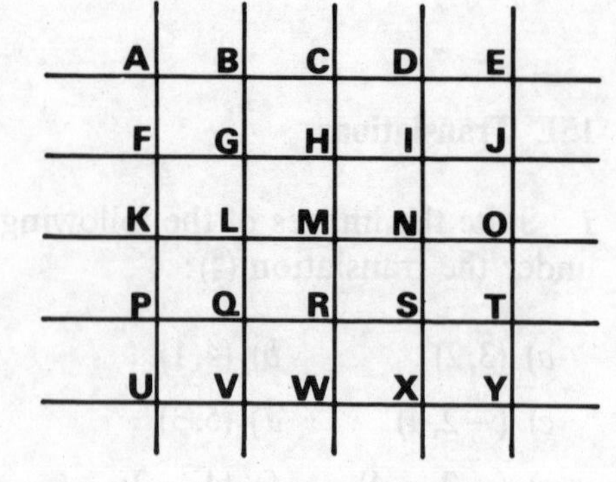

5 Give other vectors which produce the same translation as

a) $\overline{QM}$ *b)* $\overline{VM}$ *c)* $\overline{JQ}$

d) $\overline{JX}$ *e)* $\overline{PT}$

6 Name two vectors which produce each of the following translations: (e.g. for *a)* $\overline{MJ}, \overline{QN}$. Many other answers are possible.)

a) $LMRQ \rightarrow IJON$
b) $STYX \rightarrow GHML$
c) $KLQP \rightarrow NOTS$
d) $QRWV \rightarrow CDIH$
e) $DEJI \rightarrow RSXW$

7 Square $LMRQ$ undergoes translations represented by the following. Give its new position in each case.

a) $\overline{UR}$ *b)* $\overline{WO}$ *c)* $\overline{HL}$ *d)* $\overline{SN}$ *e)* $\overline{ST}+\overline{PF}$

f) $\overline{AC}+\overline{IN}$ *g)* $\overline{HG}+\overline{FK}$

8 A translation $\boldsymbol{t}$ maps the origin on to the point (2, 3). Give the points on to which the following are mapped under the same translation:

a) (1, 1) *b)* (3, 0) *c)* (−1, 2) *d)* (2, −3) *e)* (−3, −2)

Check your calculations by plotting on squared paper.

9 On graph paper draw the square S with vertices at (0, 0) (0, 1) (1, 1) and (1, 0).

If $\boldsymbol{p}$ maps (0, 0) on to (0, 2) and $\boldsymbol{q}$ maps (0, 0) on to (4, 2), draw the following squares:

a) S_1 which is S translated by $2\boldsymbol{p}$ *c)* S_3 which is S translated by $-\boldsymbol{p}$
b) S_2 which is S translated by $\boldsymbol{q}$ *d)* S_4 which is S translated by $\boldsymbol{p}-\boldsymbol{q}$

10 T is the translation $\binom{1}{1}$
S is the translation $\binom{2}{0}$
V is the translation $\binom{-3}{-1}$
O is the triangle whose vertices are (3, 2), (2, 0), (1, 2).

a) Write down the vertices of

i) $T(O)$ *ii)* $S(O)$ *iii)* $V(O)$ *iv)* $TS(O)$ *v)* $TV(O)$ *vi)* $TSV(O)$

Note $TS(O)$ means 'first S on O, then T on the result'.

b) Illustrate *a iv*), *v*) and *vi*) graphically.
c) Give a single translation equivalent to each of *a iv*), *v*) and *vi*)

11 . (For chess enthusiasts)
Consider white only. Which pieces can move in the manner represented by the vectors stated? The upper figure of the vector gives motion to the right and the lower figure motion forward. Give as many examples as you can in each case.

a) $\begin{pmatrix}7\\7\end{pmatrix}$ *b*) $\begin{pmatrix}4\\4\end{pmatrix}$ *c*) $\begin{pmatrix}-4\\-4\end{pmatrix}$ *d*) $\begin{pmatrix}1\\1\end{pmatrix}$ *e*) $\begin{pmatrix}1\\2\end{pmatrix}$ *f*) $\begin{pmatrix}2\\1\end{pmatrix}$

g) $\begin{pmatrix}-2\\-1\end{pmatrix}$ *h*) $\begin{pmatrix}3\\0\end{pmatrix}$ *i*) $\begin{pmatrix}0\\3\end{pmatrix}$ *j*) $\begin{pmatrix}2\\-2\end{pmatrix}$ *k*) $\begin{pmatrix}1\\0\end{pmatrix}$ *l*) $\begin{pmatrix}-1\\0\end{pmatrix}$

m) $\begin{pmatrix}0\\1\end{pmatrix}$ *n*) $\begin{pmatrix}0\\-1\end{pmatrix}$

* *12* A plane is to be filled with an infinite tessellation of hexagons each of side 5 cms.

a) Draw one of these hexagons, centre *O* in the middle of a sheet of paper. Surround it with six more and surround these with a ring of twelve. Taking *O* as origin and the *x*-axis parallel to one side of the hexagons, by actual measurement on your drawing find out and write down the six translations which will map the centre hexagon on to any hexagon in the first ring.
(*Note* Make as few measurements as possible. Many of the answers follow by symmetry.)
b) Write down the twelve translations which will map the centre hexagon on to any of the second ring of hexagons.

13 Trace on to a sheet of graph paper a map showing your own town and six surrounding towns or villages.
Give vectors which will map your own town on to each of these surrounding towns or villages in turn.
(Make your measurements from the centre of the town and give all distances in km to the nearest quarter of a km.)

14 On a sheet of graph paper draw to scale a diagram of a hockey, netball, rugger or soccer pitch and mark on it one possible path of the ball from the 'centre' to a successful goal, giving about a dozen 'legs' in the path.
Give the game concisely as a set of translations.

a) To what would these translations always add? Check that they do so in your particular case.
b) In an actual game, what would be the maximum possible value of the sum of all the vectors from the start to half time?

15 Which properties remain constant under a translation? (Compare 13A, question 6.)

16 Everyday Arithmetic and Off the Beaten Track

16A Everyday Arithmetic

Prices etc. stated in this exercise are reasonable at the time of going to press. The teacher should however update them where necessary.

1 A rectangular room is 4 metres by 5 metres. The centre of the floor is covered by a carpet, and there is a surround $\frac{1}{2}$ metre wide which is polished.

a) State the length and breadth of the carpet.
b) Calculate its area, giving the correct unit.
c) Find the cost of the carpet at £12 a square metre.
d) Draw a sketch of the carpet and surround, and show how the surround can be divided into rectangles which can be fitted together to give a long, narrow rectangle. State the dimensions of this rectangle.
e) What is the area of the surround, and what is the cost of polishing it at 70 pence per square metre?

2 Repeat question 1 with a room $4\frac{1}{2}$ metres square, a carpet costing £15 a square metre, and the other details the same.

3 Repeat question 2 with a surround 70 centimetres wide.

4 The walls of the room in question 1 are 2·4 metres high. They are to be papered. There is a fireplace 1·1 metres high and 1·3 metres wide, a French window 2·1 metres high and 1·1 metres wide, and a connecting door 2·1 metres high and 0·9 metres wide.

a) Find the total area of the walls, making no allowance for doors, windows, etc.
b) Find the area of the items listed: fireplace, door, etc.
c) Find the total area to be papered.
d) If the paper is sold in rolls ten metres long and 60 cm wide, find the number of rolls required. (Go to the nearest whole number above the actual requirement.)
e) Would it make a significant difference to your final answer if, to get a simple calculation, you worked all the smaller areas by rounding off the lengths to the nearest metre? Work it out and see.

5 Repeat question 4, but this time allow for a skirting board 25 cm high, and a picture rail 45 cm below the ceiling. Only the area between the skirting board and the picture rail is to be papered.

6 Windows and doors are generally recessed. The edges of the recess would have to be papered. What is an easy way of allowing for this extra paper without doing a long calculation?

7 Discuss with your parents or teacher how many hours it would take a handyman to paper the room. If the paper costs £4 a roll and the handyman's time is priced at £1.50 an hour, what would be the cost of papering the room?

8 A professional decorator would paper the room in half the time assumed in question 7. His labour would be costed at £2 an hour. Adding the cost of labour to the cost of the paper and increasing your total answer by 50% to allow for overheads, such as travelling time, wear and tear on equipment, etc., what could a professional decorator reasonably charge to paper this room?

* *9* A gardener wishes to plant roses in a rectangular bed measuring 4 metres by 3 metres. Ideally the roses should be 80 cm apart, and no rose should be nearer to the edge than 40cm. These measurements may, however, be increased or decreased by up to 10cm if necessary.

Draw a planting diagram, count the number of roses required, and estimate the cost of the roses at £1.25 each.

* *10* Repeat question 9 for a rectangular bed
a) 5 metres by 4 metres *b)* 6 metres by 4 metres.

* *11* Repeat for a circular bed of radius 3 metres.

* *12* Examine the cost of running a car, given that:

Full comprehensive insurance is £80 a year.
Tax is £50 a year.
Repairs and servicing cost £20 a month.
Tyres cost £15 each and their life is 20000 kilometres.
Petrol is 20p a litre and the car averages 12 kilometres to the litre.
Depreciation is £300 a year.
The cost of garaging is £15 a month.

Calculate
a) the total of the fixed annual costs
b) the annual cost of petrol and tyres for 25000 km/year
c) the total annual cost *d)* the total cost per km

13 *a)* What would be the cost per km
i) if the motorist ran 10 000 kilometres per year
ii) if he ran 40 000 kilometres per year?
b) If the motorist ran 200 000 kilometres per year, which costs in question 12 would need revision?

14 A typist takes a firm's mail to post. There are 128 letters second class (7p each) and 12 first class (9p each). There are also some heavier second class letters, 10 at 15p and four at 24p. There are seven parcels, four at 59p and three at 82p.

How much change would she have from two £10 notes?

15 A school party consisting of eighty boys and 120 girls under sixteen, ten boys and eight girls over sixteen and 15 teachers makes a journey to a science exhibition. They travel by rail on a special party ticket. The fare is £1 for each pupil and £2 for each accompanying adult, but free tickets are allowed for adults at the rate of one for every fifteen pupils. The remaining journey (by coach) costs 30p for each pupil and 60p for each accompanying adult.

What is the total cost of the journey? What part of the above information is irrelevant?

16 If in question 15 the teachers decide that they will pay the same amount as the boys and girls, and that the charge per head should be rounded off to the nearest five pence above, how much should each person pay? How much will be left over?

17 If the party in question 16 is insured at the cost of five pence for each boy or girl and ten pence for each teacher, what would be the total cost per head?

18 If ten mechanical diggers can construct an earth bank in four days, how long would it take eight diggers to construct the same bank?

If the ten diggers required forty waggons to keep them working to capacity, how many waggons would the eight diggers require?

If a 'day' consists of eight hours plus two hours overtime, i.e. ten hours altogether, and if it is decided to build the bank in the same number of days using eight diggers instead of ten, how many hours overtime would they have to work each day?

* *19* In question 18 list the advantages and disadvantages of using

a) ten diggers *b)* eight diggers.

20 A chef cooking for a party of ten makes enough cakes and pasties to last three days. If two members of the party are away, how long would the cakes last? Your answer should be a whole number of days.

If due to unforeseen circumstances all but two members of the party were away, how long would the cakes last?

Are your answers reasonable? Discuss.

21 A newly-married couple order bulbs for the garden of their house. Prices quoted are for quantities of 100, and smaller quantities (down to 25) are charged pro rata.

Variety	*Price in £'s*	*Quantity ordered*
Hyacinth – Lady Derby 17/18 cm	9.00	50
Daffodil – Magnificence	3.30	100
Narcissus – Havelock	2.25	50
Tulips – double early mixed	5.50	50
Tulips – Darwin, William Pitt	1.50	25
Crocus – mixed	1.75	200
Snowdrops – mixed	2.00	200
Fritillaria	1.50	50

There is $2\frac{1}{2}\%$ discount for cash with order. 80p must be added for carriage and packing, unless the order exceeds £10. Carriage and packing is then free.
They enclose a cheque with the order. For how much should they make it?

* **22** A bath is filled from two taps, one supplying hot water and one cold. The cold comes straight from the mains and runs twice as fast as the hot.
If it takes three minutes to fill the bath with both taps turned on, how long would it take to fill with *a*) the cold tap only, *b*) the hot tap only, *c*) the hot tap on full and the cold tap on half way.

23 A computer was asked, 'Which keeps the better time, a clock which loses one minute a day or a clock which doesn't go at all?' Its answer was, 'The clock that doesn't go at all'.
The computer's answer was a foolish one. On what reasoning could it have been based? Who or what was at fault?

24 A school party starts the Pennine Walk from Edale. They find that their rate of progress is as follows:

Bog	2 km per hour
Steep up	2 km in 50 minutes
Steep down	2 km „ 50 „
Rough fell	2 km „ 25 „
Good walking	2 km „ 15 „

If they leave a hostel at 0830 hours and their day's walk to the next hostel consists of 5 km of bog, 1 km steep up, 1 km steep down, 20 km rough fell and 18 km of good walking, and if they rest on an average $\frac{1}{2}$ hour after every $3\frac{1}{2}$ hours of walking, at what time would they expect to reach the next hostel?

25 In question 24, is the party likely to arrive at exactly the time you calculated? Discuss.

26 A man employs a jobbing gardener and finds that it takes him half an hour to mow the lawn, half an hour to cut the edges, one hour to weed the flower beds, three hours to cut the hedges and clean up, and two hours to do the necessary work in the vegetable garden. The hedges only need cutting every ten weeks. The gardener leaves, and he employs a second gardener whose times are as follows: 45 minutes, 45 minutes, 2 hours, five hours, and three hours. His charge per hour, however, is only 70% of the charge made by the first gardener.

Which gardener is the cheaper?

Is the cheaper necessarily the better? Discuss.

* **27** A manufacturer sells four kinds of biscuits, and the table shows the number of biscuits per packet and the wholesale price charged:

	Number of biscuits per packet	*Wholesale price per packet*
A	24	15p
B	22	18p
C	20	16p
D	18	20p

With rising prices he finds he is making insufficient profit, so to increase his profit and at the same time promote sales he decides to reduce the number of biscuits per packet by two (while keeping the size of the packet unchanged), and at the same time reduce the price by 1p.

Assuming that all four kinds sell in approximately equal quantities, will he have increased his profit?

* **28** Four children are cutting shapes from rectangular sheets of paper measuring 20 cm by 16 cm.

a) George is cutting rectangles 3 cm by 2 cm. What is the largest number he can get?

b) Susan is cutting squares of side 3 cm. What is the largest number she can get?

c) Peter is cutting circles of diameter 3 cm. What is the largest number he can get?

d) Glynis is cutting hexagons of side 1·5 cm. What is the largest number she can get?

e) Put the number of shapes obtained by Susan, Peter and Glynis in numerical order. Explain any similarities or differences.

* **29** *a*) If the paper in question 28 had been 21 cm by 16 cm, would the order in part *e*) have been different? Explain.

b) If the paper in question 28 had been very large (say 200 cm square) what order would you have expected in part *e*)?

16B Off the Beaten Track

To answer some of these questions you will need to look up facts in a book about mathematics in your school or town library; or perhaps your parents or older friends can help you. A very useful book to consult is *Number: The Language of Science* by Tobias Dantzig.

1 Ask your neighbour to think of a single figure number, not telling you what it is. Tell them to double it, then add 6, then halve it, then take away the number they first thought of. Now tell them they have 3 left. If they've done their sum correctly, they will agree. Try it again, this time telling them to add 4. The final answer should be 2. Next tell them to multiply by 3 and add 6. Then divide by 3 and take away the original number. The answer should be 2.

Can you make up similar examples for yourself? Practise them on your friends or at home. Can you see how they work?

2 Here is a method used in former times by French peasants for multiplying two numbers between 5 and 10 using their fingers. (Thumbs count as fingers also.)

To multiply 7 by 9: 7 is $5+2$, so hold two fingers up on one hand. 9 is $5+4$, so hold four fingers up on the other hand. Add the fingers that are up: $2+4 = 6$. Multiply the fingers that are down: $3 \times 1 = 3$. The answer is 63.

Representing 7
(2 up)

Representing 9
(4 up)

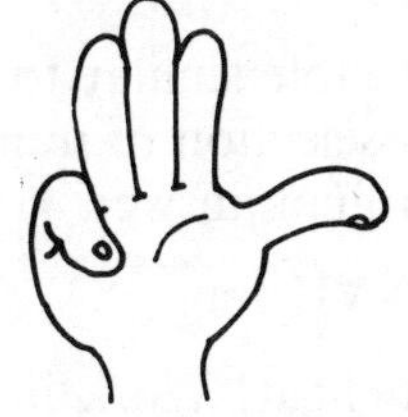

Try it yourself for 6×8, 7×7, 8×8, 8×7 and 6×7. In the last example you meet a snag. Can you overcome it? (*Hint* The 3 means 30.)

3 Name one of the pioneers of set theory, who was also an Oxford don and wrote some very imaginative books for children, which are still favourites today. When did he live? Mention two of his better known books for children.

4 Write down the Roman numerals from I to L.

5 Write down the Roman numerals from LI to C.

* *6* You will need an abacus (or a child's bead frame) to answer this question. This should have several rows of counters with ten on each. If you can't get one, draw a number of free hand sketches of a frame, and pencil in the beads as you go along. Here is a typical sketch.

Imagine you were a Roman sage and you wanted to multiply XXVII by XXIII. First set up XXVII on your abacus. Then add XXVII to it. You should then have LIV. Now add LIV to LIV and you should have CVIII. Repeat the process and get CCXVI. Do it once more and get CCCCXXXII or CDXXXII. You have now multiplied XXVII by II, IV, VIII and XVI. Tabulate your results.

XXVII × I = XXVII
XXVII × II = LIV
XXVII × IV = CVIII
XXVII × VIII = CCXVI
XXVII × XVI = CDXXXII

Now you have to select from I, II, IV, VIII and XVI to get a total of XXIII. The selection is XVI+IV+II+I.

So for the final answer you must add (on the abacus) CDXXXII, CVIII, LIV and XXVII. Adding these gives DCXXI which is the correct answer. Now try your hand on the following multiplications, in each case giving as your answer

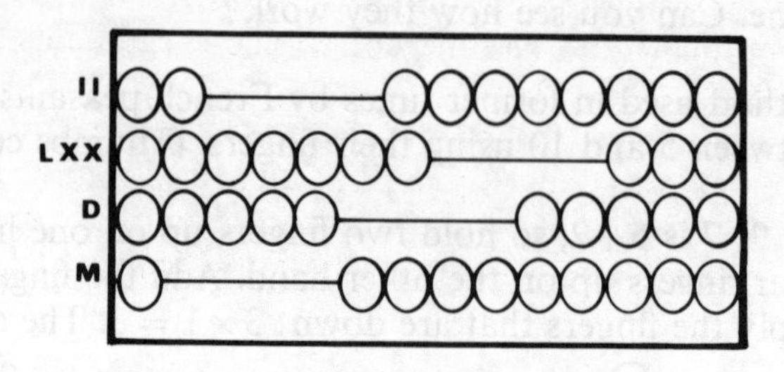

1 a table similar to the one above
2 a selection of numbers I, II, IV etc.
3 a final answer.

a) XVII times XV *b*) XLII times XIX *c*) XXXVII times XXI

Don't cheat! Work in Roman numerals using the abacus.

7 What name is given to the process described in question 6?

8 Question 6 should have given you a healthy respect for any Roman sage who could multiply together two numbers bigger than X.
Most of us, however, can multiply together two 2 figure numbers with the greatest of ease. This is because we use *positional numeration.* In this system a digit has a different meaning according to where it occurs in a number. Thus a 2 can mean 2, 20, 200, 2000 etc.
What is the meaning of 3 in the following numbers

a) 32 *b*) 3271 *c*) 163 *d*) 328?

9 Positional numeration was impossible without the 'invention' of zero. See if you can find the answers to the following questions:

a) Who first used a symbol for zero?
b) How long ago?
c) Name a race of people who would not allow the idea of zero to be used in mathematics.
d) On what grounds did they reject it?
e) What was the effect of this on the development of mathematics?
f) What is our attitude today to arguments such as they used?

* *10* Tom was given an apple to share with his younger sister June. He cut it into two, kept the 'bigger half' himself and gave the 'smaller half' to June.
'Oh, Tom, you are mean,' wept June. 'If I'd cut the apple I'd have given you the big half and kept the small half

myself.' 'What are you complaining about?' asked Tom. 'Isn't that what you've got?' In this story we have referred to 'bigger half' and 'smaller half'. But two halves should be equal. Think of half a chair, half a cake, half a bag of sweets, half a kilogram of sugar.

Are the two halves ever equal? You might think they are for the sugar, but try using a more sensitive balance and you'll soon find they aren't. Moreover, even if the weights were the same, the sizes, shapes and number of the individual grains would be different. Similar arguments apply to thirds, quarters etc.

The race of people referred to in 9 *c*) rejected the idea of fractions for this very reason. Once again the development of a part of mathematics was held up for centuries. Nowadays we say 'Even if exact fractions don't exist in practice, lets pretend they do and see where this leads us!'

Either

a) Discuss whether or not exact fractions ever exist in everyday life.

b) Can you think of one mathematical idea where exact fractions exist?

or

c) Do exact fractions exist, and if not why do we study them? Discuss.

11 In answering this question, work by yourself and do not let your neighbour see what you are doing till you've finished.

Write down a number of three different digits, the first being bigger than the third (e.g. 846, but not 648). Underneath write a second number with the same three digits in reverse order (e.g. 724 and 427). Now subtract. This will give you a third three figure number. Reverse the order of the digits of this third number and write it underneath the third number. Now add.

Next, multiply the answer by 30 (i.e. add a 0 at the end and multiply by 3). The final number contains a hidden message which you can find by decoding, using the code words:

P	E	W	T	E	R	I	R	O	N
0	1	2	3	4	5	6	7	8	9

Show your answer to your neighbour and see if he or she agrees with the message. Repeat using another three figure number. Do it several times. Explain your result.

12 This cross number is symmetrical about the lines AB and CD. Copy the grid on to squared paper and complete the shading. Add the rest of the numbers and then solve the puzzle using the given clues.

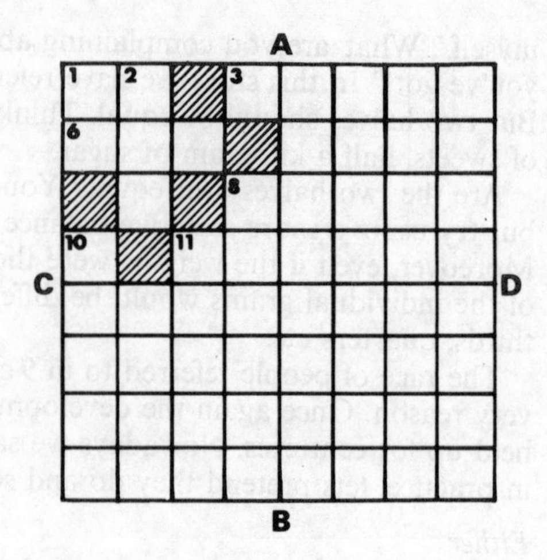

ACROSS

1 Sum of factors of 286.

3 A cake can be cut into 8 pieces. How many cakes are needed for a party of 120 people if each person has one piece?

4 $3a^2-6a+2$ when $a=4$.

6 Angle of a regular 9 sided polygon.

7 The bearing of B from A given that the bearing of A from B is 030°.

8 $\frac{1}{2}x-7=4$. Find x.

11 35·7 m + 1·82 m + 72 cm
Answer in cm.

14 Cost in pence of 12 theatre tickets at 85p each.

16 Area of a trapezium (in cm^2) if the parallel sides are 12·6 and 20·2 cm and the distance between them is 5 cm.

18 $(25\times\frac{1}{2})+(25\times 3\frac{1}{2})$.

19 Find the mean of 120, 123, 125 and 128.

21 Write 48·47 correct to 2 s.f.

22 The number of 2 cm cubes needed to make a block 12 cm by 4 cm by 6 cm.

23 The angles of a quadrilateral are 140°, 90°, 74°, x°.
Find x.

DOWN

1 1, 1, 2, 3, 5, 8, 13,?

2 1043 − 174 in base 8.

4 $2\frac{1}{7}$ as a decimal to 2 d.p.

5 $36\times 1\frac{2}{3}$

8 21·6 × 130.

9 170_{10} in base 4.

10 5, 7, 11, 19, 35,?

11 Perimeter of a rectangle $7\frac{1}{2}$ cm by 8 cm.

12 Find $\angle ABC$.

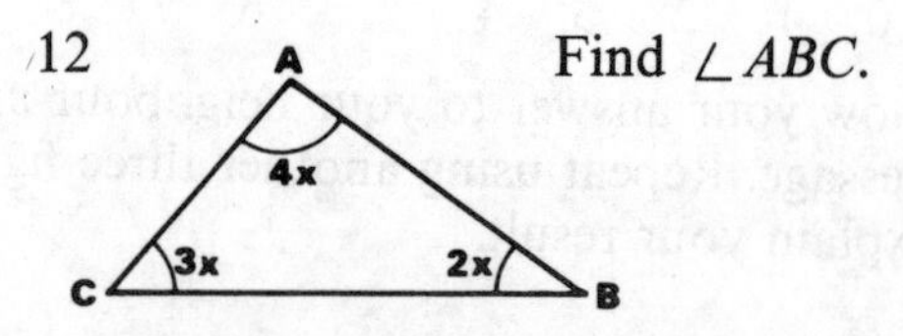

13 A man travels by car at 54 km/hr. After how many minutes has he travelled 45 km at this constant speed?

15 Cost in pence of 14 coconuts at 22p each.

17 South-west written as a three figure bearing.

18 $x^2=196$ Find x.

20 9·2 ÷ 0·2

13

12	E	C	A	L	P	A	L	A	B	E	L
11	P	A	L	A	G	R	A	N	G	E	L
10	A	Y	H	C	U	A	C	A	S	T	E
9	S	L	T	M	A	N	B	E	E	A	S
8	C	E	P	H	N	B	T	L	U	M	S
7	A	Y	R	E	A	R	R	O	C	R	U
6	L	E	V	B	A	G	R	O	L	E	R
5	C	I	C	C	A	N	O	B	I	F	E
4	G	O	S	S	E	T	T	R	D	X	I
3	X	E	M	N	R	O	L	Y	A	T	P
2	D	E	I	N	S	T	E	I	N	S	A
1	R	E	L	U	E	N	O	T	W	E	N
	1	2	3	4	5	6	7	8	9	10	11

The names of these mathematicians are hidden in the grid. Give the co-ordinates of the initial and final letters of their names.

Example Newton (11, 1) to (6, 1).

Abel	Euclid	Napier
Babbage	Euler	Newton
Boole	Fermat	Pascal
Cauchy	Fibonacci	Pythagoras
Cayley	Gossett	Russell
Descartes	Lagrange	Taylor
Einstein	Laplace	Venn

Miscellaneous Examples C

C1

1 Add another three terms to each of the following sequences:

a) 1 2 3 5 8 13
b) 1 3 6 10 15
c) 1 2 4 8 16
d) $\frac{2}{3}$ $\frac{3}{7}$ $\frac{7}{8}$ $\frac{8}{12}$

2 Write each of the following numbers as the product of prime numbers:

18 40 164 819 3978

3 A girl was given a box of chocolates on her birthday. She gave a quarter of them to her sister; $\frac{2}{5}$ of the remainder were eaten by the guests at her party. Her mother and father had two chocolates each and the girl was then left with five chocolates for herself. How many did the full box contain?

4 Write down the equations satisfied by the co-ordinates of these sets of points:

a) {1,4) (3,6) (5,8) (9,12)}
b) {(6,4) (5,5) (2,8) (1,9)}
c) {(0,0) (4,12) (6,18) (7,21)}
d) {(1,3) (3,7) (5,11) (7,15)}
e) {(1,2) (3,8) (7,20) (9,26)}

5 A ship sails from the port of Albion to Sima, a small island 4 km away on a bearing of 072° from Albion. From Sima the ship then sails 6 km on a bearing of 200° to another small island Donna. Make an accurate scale drawing of the voyage and from your drawing find how far Donna is from Albion and on what bearing the ship must sail in order to return direct to Albion.

C2

1 Plot the points $A(3,1)$ $B(5,5)$ $C(1,4)$ using the same scale on both axes.

a) Draw the perpendicular from A to BC and measure its length. *b*) Measure the length of BC. *c*) Using these two measurements, find the area of triangle ABC. *d*) Find the area of triangle ABC by 'boxing in' (see 10A, question 8*a*).

2 *a*) A solid has 9 planes of symmetry, 13 axes of symmetry and one centre of symmetry. What is it called? What is its order of symmetry?
b) Another solid has three planes of symmetry, three axes of symmetry

and one centre of symmetry. What is it called? What is its order of symmetry?
c) A plane figure has four sides, no axis of symmetry but rotational symmetry of order 2. What is it?
d) Another plane figure has four sides, one axis of symmetry but no centre of symmetry. What is it?

3 Add *a*) $3a+2b-c$ and $a-4b-2c$ *b*) $4a-b-6c$ and $5a+b-3c$

Subtract *c*) $2p-3q+6r$ from $3p+2q+r$ *d*) $p-q-r$ from $p+q-r$

If $a = 1$, $b = 2$ and $c = 3$, find the value of *e*) $3a-2b-c$ *f*) $2a^3b^2c$

4 If possible, draw networks with the following (and no other) nodes (2-nodes are ignored):

a) two 3-nodes, one 5-node and one 1-node
b) two 3-nodes
c) one 5-node and six 3-nodes
d) six 4-nodes
e), *f*), *g*), *h*) For each figure, state whether it is traversable; if not, state the number of journeys in which it can be traversed. State also the starting points.

5 Using the sieve of Eratosthenes, list the prime numbers between 800 and 850.

C3

1 If $b * c$ means 'square b and add twice c', find the value of

a) $3 * 4$ *b*) $4 * 3$ *c*) $1 * 2$ *d*) $-1 * 2$

e) $2 * (3 * 4)$ *f*) $(2 * 3) * 4$

2 A wooden block in the shape of a cube has its eight corners sawn off, leaving eight triangular faces, the sides of these triangles being about one tenth of the length of the side of the original cube. Count systematically the number of faces, edges and vertices of the resulting solid and check that it obeys Euler's relation, $F-E+V = 2$.

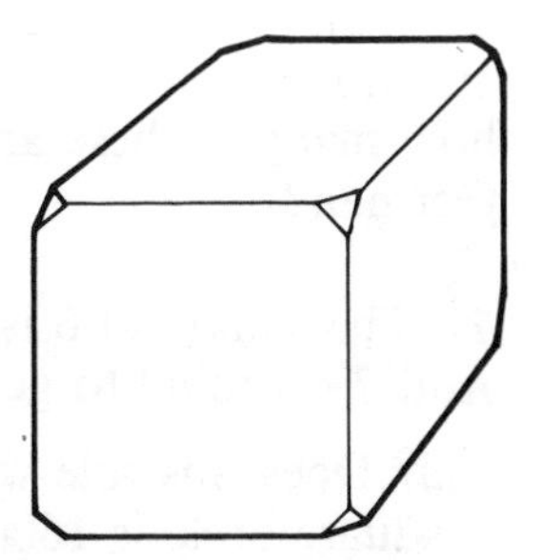

3 Solve the equations

a) $3b+5 = 2b-6$ *b*) $3b-4-3b+2 = 4b-1$ *c*) $b-6+2b+6 = 3b+1$

d) $x-2y+4 = 3x-2y+2$ *e*) $a-4+2a-3 = 3a+5+a$

If no solution is possible, say why.

4 Find the area of a regular hexagon of side 4 cm. Use any method you like.

5 A class consists of 15 boys and 17 girls. In an examination the average marks for the boys and girls, totalled separately, were as follows:

	Maths	*French*	*English*	*Science*	*Art*
Boys	51	48	42	56	54
Girls	48	51	55	41	54

What was the average mark for the whole examination for *a*) the boys *b*) the girls *c*) the whole class?
d) Give the average mark for the whole class in Maths and English.

C4

1 $A = \{x : x \text{ is an integer and } 0 < x < 200\}$
$B = \{x : x \text{ is an odd integer and } 0 < x < 200\}$
$C = \{x : x \text{ is a multiple of 3 and } 0 < x < 200\}$
$D = \{x^2 : x \text{ is an integer}\}$
$E = \{x^3 : x \text{ is an integer}\}$

Write down *a*) $n(C)$ *b*) $E \cap A$ written out in full *c*) $E \cap A$ expressed in words *d*) $n(B \cap D)$ *e*) $n(C \cap E)$

* **2** Subtract 3014_5 from 3014_6, multiply the result by 123_7 and give the answer in base 8. (You may find it helps to change everything to base 10.)

3 In a youth hostel, two thirds of the hostellers are British, one-sixth are German and one-eighth are French. The hostel has accommodation for 50, and this number is never exceeded. How many hostellers are there altogether, how many of these are British, and how many are neither British, French nor German?

4 The square of 6 is 36. Add $6+7$ to get 49. This is the square of 7. Add $7+8$ to 49 to get 64. This is the square of 8.

a) Does this rule work for finding the square of 41, knowing that the square of 40 is 1600?
b) Does it work for any pair of numbers differing by 1?
c) Can you adapt it to get the square of 52, being given that the square of 50 is 2500? You can of course apply the old rule twice over, but can you find a new rule which gives the answer in a single step?
d) Use your new rule to find the square of 122, given that the square of 120 is 14 400.

5 In a time of inflation, a market research team kept a record of the change in prices of 100 different items of food commonly found on the shelves of a supermarket.
The table shows the percentage increase in price over a year:

Number of articles	*% increase in a year*
4	20% to 25%
8	15% to 20%
16	10% to 15%
50	5% to 10%
20	0 to 5%
2	decreased by 2%
100	

Represent these figures *a*) on a bar chart *b*) on a pie chart. Which of these two charts gives the clearer picture of the degree of inflation?

* What is the average increase in price of these 100 articles? Does this 'average increase in price' give the increased cost to the consumer, or would another average be necessary? Discuss.

C5

1 Change the given fractions to decimals, put the decimals in order of size and thus put the fractions in order of size:

$\frac{2}{3}$ $\frac{3}{4}$ $\frac{14}{17}$ $\frac{16}{23}$ $\frac{27}{37}$

2 State the size of the angles marked with letters. Give brief reasons for your answers:

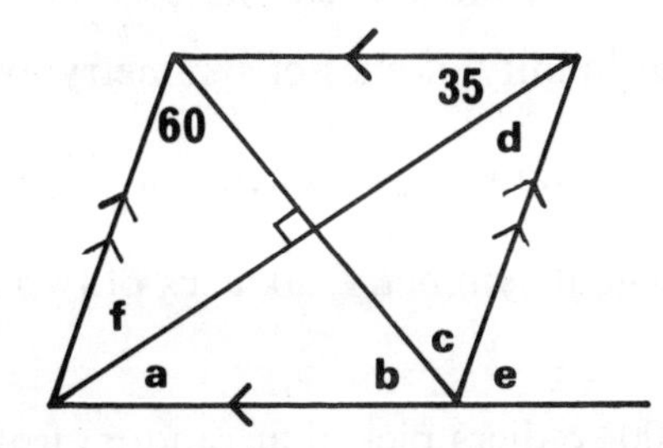

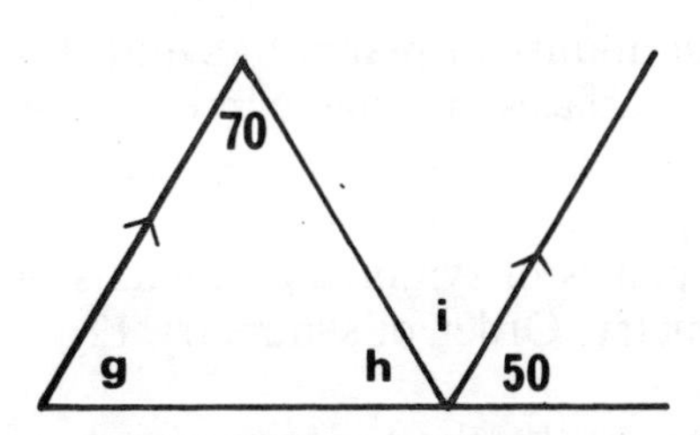

Note The figures are not drawn accurately. You must calculate the angles, not measure them.

3 How many axes of symmetry can you draw in a regular polygon of

a) 4 sides *b*) 5 sides *c*) 6 sides *d*) 7 sides *e*) *n* sides?

Draw figures to illustrate your answer to numbers *a*) to *d*). In *e*), state where the axes are drawn when *n* is even and where they are drawn if *n* is odd.

4 Fill in the missing signs:

a) 3 ____ $(4-5) = 4$ *b*) $4 \times (-4) \times ($____ $4) = 64$

c) $5-(3$ ____ $12) =$ ____ 14 *d*) $(4$ ____ $3)$ ____ $6 = -5$

5 A class was challenged by their teacher to draw a network that could not be coloured with four colours. Ian and Fiona working together, triumphantly produced the following network, the numbers 1 to 4 representing different colours. Where did they go wrong? Draw the network and number it correctly.

1
3 2 3 2
2 4 1 4 3
3 2 3 2 1 2
2 1 3 4 1 3 4
3 2 3 4 1 3
2 1 3 4 1 2
3 4 2 3 4 3 4

C6

1 Write down the equations of the lines joining the following sets of points:

a) (3, 4) (2, 5) (1, 6) *b*) (3, 6) (4, 8) (5, 10) *c*) (3, 7) (4, 9) (5, 11)

d) (3, 5) (4, 6) (6, 8) *e*) (3, 4) (5, 5) (7, 6)

2 State whether the following equations have solutions in the sets indicated. If the answer is YES, give the solution(s).

a) $x+3 = 2$ (integers, positive integers)
b) $2x+3 = 2$ (integers, negative integers)
c) $3x+6 = 2x+6$ (positive integers, integers)
d) $5x-7 = 2x+3$ (positive integers, negative integers)

3 Name the following solids which have the symmetry, etc. stated:

a) An infinity of planes of symmetry, an infinity of axes of symmetry, two plane surfaces and one curved surface.
b) 4 planes of symmetry, 4 axes of 3-fold symmetry, symmetry of order 12.
c) 6 planes of symmetry, one axis of six-fold symmetry.
d) 7 planes of symmetry, one axis of six-fold symmetry, six axes of two-fold symmetry. Order of symmetry 12.

4 Here are the chest measurements of 100 soldiers picked at random from a line regiment:

Chest measurement in cm	*No. of soldiers*
84– 88	2
88– 92	7
92– 96	25
96–100	45
100–104	17
104–108	4
	100

Represent these measurements as a bar chart and state the modal chest measurement.

Taking the whole group as having the measurement given by the mid-point of the group (i.e. taking the first group as all having 86 cm) calculate the mean chest measurement for the 100 soldiers. What is the median measurement?

5 Plot the points (2, 4) (4, 8) (10, 10) (12, 5) (12, 2) (7, 1) and join them up in that order to form an irregular closed hexagon.

Find the area of this hexagon.